Fire Suppression Substitutes and Alternatives to Halon for U.S. Navy Applications

Committee on Assessment of Fire Suppression Substitutes and Alternatives to Halon
Naval Studies Board
Commission on Physical Sciences, Mathematics, and Applications
National Research Council

National Academy Press
Washington, D.C. 1997

NOTICE: The project that is the subject of this report was approved by the Governing Board of the National Research Council, whose members are drawn from the councils of the National Academy of Sciences, the National Academy of Engineering, and the Institute of Medicine. The members of the committee responsible for the report were chosen for their special competences and with regard for appropriate balance.

This report has been reviewed by a group other than the authors according to procedures approved by a Report Review Committee consisting of members of the National Academy of Sciences, the National Academy of Engineering, and the Institute of Medicine.

The National Academy of Sciences is a private, nonprofit, self-perpetuating society of distinguished scholars engaged in scientific and engineering research, dedicated to the furtherance of science and technology and to their use for the general welfare. Upon the authority of the charter granted to it by the Congress in 1863, the Academy has a mandate that requires it to advise the federal government on scientific and technical matters. Dr. Bruce Alberts is president of the National Academy of Sciences.

The National Academy of Engineering was established in 1964, under the charter of the National Academy of Sciences, as a parallel organization of outstanding engineers. It is autonomous in its administration and in the selection of its members, sharing with the National Academy of Sciences the responsibility for advising the federal government. The National Academy of Engineering also sponsors engineering programs aimed at meeting national needs, encourages education and research, and recognizes the superior achievements of engineers. Dr. William A. Wulf is interim president of the National Academy of Engineering.

The Institute of Medicine was established in 1970 by the National Academy of Sciences to secure the services of eminent members of appropriate professions in the examination of policy matters pertaining to the health of the public. The Institute acts under the responsibility given to the National Academy of Sciences by its congressional charter to be an adviser to the federal government and, upon its own initiative, to identify issues of medical care, research, and education. Dr. Kenneth I. Shine is president of the Institute of Medicine.

The National Research Council was organized by the National Academy of Sciences in 1916 to associate the broad community of science and technology with the Academy's purposes of furthering knowledge and advising the federal government. Functioning in accordance with general policies determined by the Academy, the Council has become the principal operating agency of both the National Academy of Sciences and the National Academy of Engineering in providing services to the government, the public, and the scientific and engineering communities. The Council is administered jointly by both Academies and the Institute of Medicine. Dr. Bruce Alberts and Dr. William A. Wulf are chairman and interim vice chairman, respectively, of the National Research Council.

This work was performed under Department of the Navy Grant N00014-96-1-0220 issued by the Office of Naval Research under contract authority NR 201-124. However, the content does not necessarily reflect the position or the policy of the Department of the Navy or the government, and no official endorsement should be inferred.

International Standard Book Number 0-309-05782-5

Copies available from:
Naval Studies Board
National Research Council
2101 Constitution Avenue, N.W.
Washington, D.C. 20418

COVER: Fuel oil fire in U.S. Coast Guard full-scale shipboard fire test facility aboard the *State of Maine*, docked in Mobile, Alabama. Photo courtesy of Richard L. Hansen, U.S. Coast Guard R&D Center, Groton, Conn.

Printed in the United States of America

COMMITTEE ON ASSESSMENT OF FIRE SUPPRESSION SUBSTITUTES AND ALTERNATIVES TO HALON

DAVID W. McCALL, AT&T Bell Laboratories (retired), *Chair*
RONALD L. ALPERT, Factory Mutual Research Corporation
JAMES G. ANDERSON, Harvard University
PHILLIP J. DINENNO, Hughes Associates, Inc.
DAVID A. DIXON, Pacific Northwest Laboratory
RICHARD E. FERNANDEZ, DuPont Fluoroproducts
ROBERT H. GORMLEY, RADM, USN (retired), The Oceanus Company
MALCOLM K. KO, Atmospheric and Environmental Research, Inc.
RUDOLPH PARISER, R. Pariser & Co., Inc.
AKKIHEBAL R. RAVISHANKARA, National Oceanic and Atmospheric Administration
REVA RUBENSTEIN, U.S. Environmental Protection Agency
GEORGE M. RUSCH, Allied Signal, Inc.
ROBERT E. TAPSCOTT, University of New Mexico
TIMOTHY J. WALLINGTON, Ford Motor Company
DONALD J. WUEBBLES, University of Illinois at Urbana-Champaign

Navy Liaison Representative

Ronald DeMarco, Office of Naval Research

Consultants

Sidney G. Reed, Jr.
James G. Wilson

Staff

Peter W. Rooney, Program Officer
Ronald D. Taylor, Director, Naval Studies Board
Susan G. Campbell, Administrative Assistant
Mary G. Gordon, Information Officer
Christopher A. Hanna, Project Assistant

NAVAL STUDIES BOARD

DAVID R. HEEBNER, Science Applications International Corporation (retired), *Chair*
GEORGE M. WHITESIDES, Harvard University, *Vice Chair*
ALBERT J. BACIOCCO, JR., The Baciocco Group, Inc.
ALAN BERMAN, Applied Research Laboratory, Pennsylvania State University
NORMAN E. BETAQUE, Logistics Management Institute
NORVAL L. BROOME, Mitre Corporation
GERALD A. CANN, Raytheon Company
SEYMOUR J. DEITCHMAN, Chevy Chase, Maryland, *Special Advisor*
ANTHONY J. DeMARIA, DeMaria ElectroOptics Systems, Inc.
JOHN F. EGAN, Lockheed Martin Corporation
ROBERT HUMMEL, Courant Institute of Mathematical Sciences, New York University
DAVID W. McCALL, AT&T Bell Laboratories (retired)
ROBERT J. MURRAY, Center for Naval Analyses
ROBERT B. OAKLEY, National Defense University
WILLIAM J. PHILLIPS, Northstar Associates, Inc.
MARA G. PRENTISS, Jefferson Laboratory, Harvard University
HERBERT RABIN, University of Maryland
JULIE JCH RYAN, Booz, Allen and Hamilton
HARRISON SHULL, Naval Postgraduate School (retired)
KEITH A. SMITH, U.S. Marine Corps (retired)
ROBERT C. SPINDEL, Applied Physics Laboratory, University of Washington
DAVID L. STANFORD, Science Applications International Corporation
H. GREGORY TORNATORE, Applied Physics Laboratory, Johns Hopkins University
J. PACE VANDEVENDER, Prosperity Institute
VINCENT VITTO, Lincoln Laboratory, Massachusetts Institute of Technology
BRUCE WALD, Arlington Education Consultants

Navy Liaison Representatives

Paul G. Blatch, Office of the Chief of Naval Operations, N911T1
Ronald N. Kostoff, Office of Naval Research
Ronald D. Taylor, Director
Peter W. Rooney, Program Officer
Susan G. Campbell, Administrative Assistant
Mary G. Gordon, Information Officer
Christopher A. Hanna, Project Assistant

COMMISSION ON PHYSICAL SCIENCES, MATHEMATICS, AND APPLICATIONS

Preface

On January 1, 1994, under the latest revision of the Montreal Protocol on Substances That Deplete the Ozone Layer, the domestic production of halons, widely used by the civilian, commercial, and military communities as a fire and explosion suppressant, was banned. The ban did not prohibit the use by the Navy of existing halon stocks, which may be adequate for projected use well into the next century at present levels of consumption. There is some concern, however, that potential future restrictions may limit the use of existing supplies.

The Navy is critically dependent on halon as a fire and explosion suppressant aboard existing aircraft and surface ships. Therefore, the Navy is faced with the following question: In view of the ban on halon production and the possibility of future restrictions on the use of existing stocks, should the research effort on halon substitutes be placed at a high priority or at a low priority while transferring emphasis to the search for alternative (non-halon-like) approaches? To answer that question, the Office of Naval Research requested the assistance of the Naval Studies Board.

Accordingly, the Committee on Assessment of Fire Suppression Substitutes and Alternatives to Halon was established under the auspices of the Naval Studies Board and charged to examine the following issues and provide advice to the Navy regarding its research on fire suppression systems and possible replacements for halons:

1. Beginning with a thorough understanding of the mechanism whereby halon adversely influences the ozone layer, and possibly contributes to global warming, assess the research effort directed at finding a substitute for halon as a fire and explosion suppressant, and evaluate any potential substitute for its (a) effectiveness compared to halon, (b) toxicity as a pure compound and after exposure to fire, (c) stability of the compound (e.g., thermal, material compatibility), (d) ozone depletion potential, and, if possible, global warming potential, and (e) impact on the firefighting system (e.g., weight and volume competitiveness).
2. Based on the results of Task 1, assess the potential for finding a drop-in replacement for halon, and identify the most promising areas of research.
3. Time and resources permitting, address the issue of comparable alternative approaches to fire and explosion suppression aboard military platforms.

The committee conducted a study of roughly eight months' duration beginning in April 1996 and including the following meetings and site visits:

- *April 9-10, 1996, in Washington D.C.* Held organizational meeting; received navy briefs.
- *June 10-11, 1996, in Norfolk, Virginia.* Toured halon installations aboard the USS *Kearsarge*, an amphibious assault ship. Received briefings from Naval Sea Systems Command personnel.
- *July 15-16, 1996, in Mobile, Alabama.* Toured U.S. Navy shipboard full-scale fire test facility aboard the *ex*-USS *Shadwell* and observed a full-scale test of an FM-200™-based fire suppression system in the U.S. Coast Guard shipboard full-scale fire test facility aboard the *State of Maine*. Received briefings from Naval Research Laboratory personnel and contractors and U.S. Coast Guard personnel and contractors.
- *August 14-15, 1996, in San Diego, California.* Toured North Island Naval Base and examined halon systems on U.S. Navy fixed-wing and rotary-wing aircraft. Received briefings on halon and inert gas generator systems from Naval Air Systems Command personnel and contractors.
- *November 11-12, 1996, in Washington, D.C.* Final meeting.

The resulting report represents the committee's consensus view on the issues posed in the charge.

Acknowledgments

The Committee on Assessment of Fire Suppression Substitutes and Alternatives to Halon wishes to express its gratitude to the many individuals who provided valuable information and support during the course of the study. Mr. James Homan and his staff in the Halon R&D Program in the Naval Air Systems Command, and Mr. Robert Darwin and Mr. Joel Krinsky and their staffs in the Fire Protection Division and the HVAC Submarine Life Support Division, respectively, in the Naval Sea Systems Command were exceptionally helpful. Likewise Dr. Ronald Sheinson, Head of the Combustion Dynamics Section of the Chemistry Division at the Naval Research Laboratory, and Dr. Richard Gann, Chief of the Fire Science Division of the Building and Fire Research Laboratory at the National Institute of Standards and Technology, and their staffs provided extremely valuable technical data.

The committee wishes to thank the commander and crew of the USS *Kearsarge* , a modem amphibious assault vessel (LHD-3) based in Norfolk, Virginia. The committee toured this vessel and inspected the halon installations on board. The ship's procedures with regard to the halon systems were described in detail and the committee was impressed with the logic embodied in the Navy's practices.

The committee wishes to thank the commander, crew, and staff of the *ex*-USS *Shadwell*, site of the shipboard full-scale fire test facility operated by the Naval Research Laboratory, and the staff of the *State of Maine*, site of the shipboard full-scale fire test facility operated by the U.S. Coast Guard R&D Center. Both vessels are docked in Mobile, Alabama. The committee was able to witness a full-scale test of an FM-200 TM-based fire suppression system aboard the *State of Maine*.

Finally, the committee wishes to thank the men and women of the North Island Naval Air Depot for graciously hosting the committee on its tour of the aviation maintenance facility there. The committee received informative briefings from the staff of the Naval Air Systems Command and the Naval Air Warfare Center in Lakehurst, New Jersey.

Contents

	Executive Summary	1
1	Introduction	3
	References	5
2	Assessment of Research on Alternatives for Halon	6
	Chemical and Physical Considerations in Evaluating Alternative Fire Suppression Agents	7
	Toxicology Issues	15
	Brief Overview of Studies of Halon Substitutes and Mechanisms	19
	Findings	19
	References	20
3	Atmospheric Chemistry and Evaluation of Environmental Effects of Fire Suppressants	22
	Reduction of Zone—Mechanism and Effects	22
	Atmospheric Lifetime, Ozone Depletion Potential, and Global Warming Potential as Indicators of Environmental Impact	26
	Other Environmental Effects	35
	References	42
4	Navy-Specific Issues	45
	Ship Systems	46
	Aircraft Systems	50
	Navy Halon Inventory	53
	Retrofitting Non-Halon Systems in Existing Ships and Aircraft	56
	Summary, Findings, and Recommendations	61
	Appendixes	
A	Halon Use in the Navy	63
	Why Halon? A Historical Overview	63
	Ship Fire Extinguishing Systems	65
	Aircraft Fire Extinguishing Systems	73

B	Regulation of Halon and Halon Replacements	75
	U.S. Regulation of Halons and Halon Substitutes	76
	Alternatives to use of Halons	78
	Halon Banking	79
	References	79
C	Stability and Materials Compatibility of Candidate Replacements for Halon	80
	Storage Stability	80
	Effects of Halon-Like Alternative Agents on Organic Materials and Metals	82
	References	84
D	Water Mist Fire Suppression Technology	85
	Theoretical and Design Considerations	86
	Recent Technology Advances	86
	Experimental Evaluation of Water Systems	88
	Current Status of Water Mist Systems	92
	References	93
E	Glossary	96

Executive Summary

Halons are recognized as ideal fire extinguishing agents, particularly for fighting fires caused by flammable liquids and explosive gases. They are highly effective in extinguishing fires in minimal time, are non-corrosive, and when deployed at recommended volume densities are non-toxic. Because of these characteristics, halons are widely used on board Navy ships and in aircraft.

Unfortunately, halons contribute to the depletion of Earth's stratospheric ozone layer, which results in higher levels of ultraviolet radiation at Earth's surface and thus gives rise to serious health effects. Halons are chemically related to chlorofluorocarbons (CFCs) that were earlier shown to degrade the ozone layer and consequently were banned. Although as compared with CFCs the quantities of halons released are much smaller, the halons are more active in destroying ozone and their effect is significant.

Under an Executive Order effective in January 1994, halons may no longer be manufactured in the United States. But because of the difficulty of finding a suitable fire extinguishing substitute for halons 1301 and 1211, the military services are permitted to use these chemicals for mission-critical purposes, such as fire fighting, in existing platforms (ships, aircraft, weapons, vehicles) until the current halon supply or "bank" is exhausted.

This study was undertaken by the Committee on Assessment of Fire Suppression Substitutes and Alternatives to Halon to assess research in the science and engineering relevant to identification and evaluation of alternative agents that could be developed to replace halons, including the likelihood of the existence of virtually identical agents. For alternative agents that have been identified, the committee also assessed efforts to quantify the penalties in added weight and volume of storage and distribution systems required. The study also addressed non-halon-like extinguishing systems as possible alternatives.

APPROACH AND SCOPE OF STUDY

The study involved a review of the extensive research and engineering literature covering the physical and chemical processes active in flames and involved in flame extinguishment. Information on test methodology, including agent toxicology, storage stability, and extinguishment effectiveness, was collected. Synthetic chemical schemes for the manufacture of halon alternatives were reviewed. Published lists of prospective halon replacements were evaluated. Atmospheric chemistry relevant to ozone depletion was examined in the light of recent work, and the possibility of secondary environmental effects was studied. Ozone depletion potential (ODP), a useful metric found in regulatory legislation, was described and tabulated for alternative agent candidates. Global warming potential (GWP), a measure of agent effect on climate, was treated similarly. The history of halon regulation and the likelihood of important modifications were also discussed.

In addition, information concerning Navy-specific aspects of halon replacement was assembled. The nature of Navy fires, existing extinguishing systems, and hardware configurations were reviewed, with Navy guidance, for ships and aircraft. Alternatives to halon that are under consideration for new-construction platforms were examined (HFC-227ea for non-machinery spaces on ships and HFC-125 for engine bays on aircraft). Non-halon-like methods were also reviewed (water mist systems for shipboard machinery spaces and inert gas generators for aircraft dry bay and engine bay applications). The potential for retrofit of those four alternatives on existing platforms was examined. Finally, the projected evolution of the Navy halon inventory was studied.

With these various aspects of the study carried through, the committee was able to take positions on the first two elements of the statement of task (see preface, p. vii) and to comment knowledgeably on the third.

KEY FINDINGS

- After reviewing research, development, toxicology, and engineering activities directed toward finding alternative and replacement agents for halon 1301 and halon 1211, the committee finds that in this context, the relevant aspects of the problem are being studied effectively and a comprehensive body of scientific and engineering knowledge is being developed, and the committee has identified no obvious gaps in these important efforts.
- It is unlikely that a drop-in replacement agent will be discovered that will exhibit all of the beneficial properties of halon 1301 and not also exhibit a significant environmental impact.
- Effective alternative chemical agents have been identified by the Navy and are currently being incorporated into the design of new ships and aircraft. There is a weight and volume penalty associated with these agents relative to halons, but the impact can be minimized if use of these agents is incorporated into initial platform design. Further, retrofit of these agents into existing naval platforms is technically feasible in most cases.
- In addition to the chemical replacement agents, promising alternative fire extinguishing systems such as water mist systems and inert gas generators are under consideration by the Navy for some applications. These systems are being incorporated into new-design naval platforms.

CONTENTS OF THIS REPORT

This study addresses various science and engineering aspects of the agents and methods that are being considered as substitutes for halons and halon systems used as fire suppressants by the U.S. Navy in ships and aircraft. Chapter 1 points out the Navy's ongoing need for effective fire suppression and outlines the set of extensive requirements that candidate replacements for halons must satisfy. Chapter 2 describes research on flames, mechanisms of flame suppression, methodology of agent evaluation including suppression efficiency and toxicology, manufacturability, and aspects of the search for alternative agents. Atmospheric chemistry phenomena relating to an alternative agent's capacity to deplete ozone, contribute to global warming, or give rise to surface pollution through the accumulation of noxious reaction products are covered in Chapter 3. Chapter 4 describes Navy-specific aspects of halon replacement in ships and aircraft, suggests possible courses of action, and provides the committee's preferred alternative. The appendixes give details as appropriate.

1

Introduction

"Fire is both the servant of mankind and its destructive demon," states John Lyons in his illuminating book on the subject, written for the general reader.[1] On ships and aircraft, where escape is dangerous or impossible, fire carries a special terror. The terror is compounded for warcraft, which carry large quantifies of highly flammable fuel and an arsenal of explosives and their propellants, and are required to operate in an environment that is hostile by design of an adversary. It is evident that the U.S. Navy must have at its disposal the most effective means of fighting fires that technology can supply and that also satisfy other necessary requirements.

Halons are a class of halogenated hydrocarbons that are highly effective in suppressing combustion and that, accordingly, are widely deployed on the ships and aircraft of the U.S. Navy.* Anyone who has observed an inferno of spewing fuel oil extinguished in seconds by halon can appreciate the Navy' s strong reliance on this method of protection. Beyond efficacy, halons exhibit other properties that make them ideal fire suppressants in a variety of applications. These properties include ease of distribution in obstructed spaces, low toxicity, and storage stability. Unfortunately, like the chlorofluorocarbons, halons have been identified as agents of stratospheric ozone depletion, and their domestic manufacture was terminated in accordance with international treaty and U.S. law.[2,3,4] In this study the Committee on Assessment of Fire Suppression Substitutes and Alternatives to Halon addresses various scientific and engineering aspects of the agents and methods that are being considered as substitutes for halons and halon systems.

The Navy currently has a considerable investment in halon systems, and it would be highly desirable to identify alternative agents that can be substituted for halon in existing hardware and pose no environmental threat. It is a great challenge to find a material that matches the necessary properties of halon 1301 (chemical, physical, and toxicological) to produce a "drop-in" replacement. Once a candidate halon alternative has been identified, it must be tested under the following demanding set of criteria:

1. Fire suppression effectiveness for a flooding agent, usually as measured by a cup burner, which is a laboratory-scale test that gives relative performance in terms of vapor concentration required for extinguishing a hydrocarbon flame;
2. Capability for distribution through the protected space via pipes and nozzles in a few seconds;
3. Toxicological test protocols designed to ensure that exposure of Navy personnel will not be harmful;
4. Activity in destroying atmospheric ozone (as measured by the ozone depletion potential (ODP) metric);
5. Effect on climate (as measured by the global warming potential (GWP) metric);
6. Environmental consequences of decomposition products following release;
7. Storage stability; and
8. Compatibility with materials (metals, elastomers, lubricants) contacted in storage and distribution hardware.

To be viable, an agent must satisfy all these tests. Although the test and modeling methodologies are well developed, it is still necessary to do full-scale testing to fully qualify an alternative.

* The Navy relies mainly on halon 1301 (CF_3Br) and uses smaller quantities of halon 1211 (CF_2ClBr). For purposes of comparison in evaluating alternatives to halons, halon 1301 is the type referred to more frequently in this report.

Research on flames and methods for extinguishing flames has been active in recent years, giving rise to a large literature. Theory, modeling, and direct observation of flames have attracted much attention. Flame reactions involve fuel (usually hydrocarbons in the context of this report) and oxygen. Flame reactions occur only at high temperatures (>800 K), and they are mediated by critical "radicals," the most important of which is the hydrogen atom. Therefore, the strategy for extinguishment is to cool the reaction mixture and to introduce chemical entities that will remove hydrogen radicals (e.g., bromine or, to a lesser extent, chlorine). Halon does both. Progress in the understanding of flames and their extinguishment has been quite encouraging, but it is still necessary to test extinguishing agents in full-scale conditions.

Over the past decade, a systematic, broadly based search has been implemented for halon alternatives, covering many classes of compounds that are candidate alternatives or might shed light on chemical mechanisms involved. This is an active and well-directed community. The most obvious candidates for replacement of halons are perfluorocarbons (PFCs) and hydrofluorocarbons (HFCs). These chemicals can quench flames by cooling. In addition, hydrochlorofluorocarbons (HCFCs) are under consideration; for HCFCs it is hoped that the chlorine atom will contribute to extinguishing the flame by removal of the radical, but that the molecule will not survive long enough in the atmosphere to reach the stratosphere (where it could threaten the ozone layer). Although they are difficult to synthesize, a selection of these classes of compounds are available commercially.

Toxicology is a key aspect of halon replacements. Although the Navy evacuates personnel from spaces to be flooded with halon 1301, there is always the possibility of accidental discharge, and the hazard must be carefully assessed. The toxicology of halon alternatives has been studied, and protocols for agent testing and use have been delineated.

In the evaluation of candidate alternatives it is necessary to characterize the atmospheric chemistry of each compound, and this is a productive field of research. Reactions in the lower atmosphere are important in determining a compound's lifetime and the probability of its reaching the stratosphere. The chemistry of ozone depletion has been well documented. These complex issues have been integrated successfully under the concept of ozone depletion potential (ODP), a useful metric that has been adopted in the U.S. Clean Air Act. Because many of the compounds under consideration have long atmospheric lifetimes and strongly absorb infrared radiation, they are expected to contribute to global warming. Although at this time there are no restrictions based on global warming potential (GWP), the possibility of future restrictions should be factored into the selection of any alternative agent or system. It is further necessary to prove that agent decomposition products will not give rise to ecological problems. Here again research has been active.

It is sobering to consider the extensive nature of the requirements—beyond flame supression capability—that a halon replacement must satisfy. It must be storable as a liquid (to conserve space) but must vaporize quickly to a gas on release (to fill an obstructed space and to act on the flame, within seconds). Of course, the ozone depletion potential must be acceptably low. In addition, the toxicology, storage stability, materials compatibility, and environmental consequences of decomposition products following release of the agent must be acceptable. Each of the requirements must be met.

Currently, the Navy has a considerable supply of halon (in the "bank"), as allowed under regulation, and it is projected that the bank is sufficient to protect existing ships and aircraft until their retirement from service. New-design platforms will be protected by non-halon systems. Given the history of increasingly strict environmental regulation, however, there is concern that pressure will build to destroy the existing halon bank set aside for military uses in order to preclude its eventual release. While this possibility seems remote at the present time, it is prudent for the Navy to prepare for such an outcome by identifying environmentally acceptable alternative agents and investigating systems changes that will be required.

If it becomes necessary to use halon replacements in existing platforms, it will be desirable to have identified alternative agents that can be used with minimal modification of existing hardware. Currently, there are no known alternative agents that can be substituted for halon 1301 in existing equipment without modification. There are, however, alternative agents that can be deployed in existing systems if modifications, such as increased storage capacity, are made. Thus, although there is no known "drop-in" replacement, there may be replacement agents available that can satisfy the Navy's requirements, including acceptable cost of retrofit.

In addition to consideration of alternative agents for halons, other methods of fire suppression have been examined in the context of shipboard and aircraft use. New methods (e.g. inert gas generators and advanced water mist systems) appear promising for use in some situations now protected with halon.

Replacement of halons in Navy applications is a demanding task. Fortunately, there is an abundance of active centers of research in relevant fields in the United States.

REFERENCES

1. J.W. Lyons, *Fire*, Scientific American Library, W.H. Freeman, New York (1985).
2. United Nations Environment Programme, *Montreal Protocol on Substances That Deplete the Ozone Layer* , Nairobi, Kenya (1987).
3. United Nations Environment Programme, *Report on the Fourth Meeting of the Parties to the Montreal Protocol on Substances That Deplete the Ozone Layer*, Copenhagen, Denmark, and Nairobi, Kenya (1992).
4. Clean Air Act, *U.S. Code*, Vol. 42, Title VI (1990).

2

Assessment of Research on Alternatives for Halon

The Montreal Protocol and its amendments have led to bans on the production of halogenated alkanes that can cause depletion of stratospheric ozone. Thus the very effective fire- and explosion-suppression compounds, the halons, are being phased out, as well as the chlorofluorocarbons (CFCs) so prevalently used in refrigeration and cleaning systems. The use of halons as fire and explosion suppressants is well established, and their superior performance in these applications is supported by a wealth of documentation dating from before 1950. A comprehensive review of the existing literature is not attempted here, but the reader is referred to six reviews that summarize this topic and that contain more than a thousand references directly relevant to halons.[1,2,3,4,5,6]

The use of halons for fire suppression in military systems was the result of a study by the U.S. Army in the late 1940s,[7] and no searches for new compounds had been conducted since then until the Montreal Protocol was signed. The halon-based fire suppression systems first used in aircraft and ships were CCl_4 (halon 104) and then CH_3Br (halon 1001). A shortcoming of these original systems was their toxicity in manned spaces.

Research on halons has been actively pursued since the 1940s at least, and halons 1301 and 1211, which are currently the most widely used total flooding and streaming agents, respectively, were selected from among a wide variety of candidate agents based on the following criteria: demonstrated efficacy in suppressing combustion, acceptable toxicological properties, availability in volume at acceptable cost, storage stability for at least 5 years in any climate, materials compatibility, low electrical conductivity, and acceptable activity of the compound or its combustion products (chemicals and/or particulates formed in the flame during extinguishment that may be toxic, corrosive, or both) in terms of low residue and low corrosion of metals. These are the same requirements that were noted in 1950 and 1951 in Air Force and Corps of Engineers documents.

Based on the suggestion by Molina and Rowland[8] in 1974 that the presence of chlorinated halocarbons in the stratosphere may catalyze the destruction of stratospheric ozone, a new requirement was added, namely, that any agent must have acceptable environmental characteristics as determined by ozone depletion potential (ODP). Halons 1301 and 1211 both have a relatively high ODP, and domestic manufacture has been terminated.

The fire fighting community in general, and the U.S. Navy in particular, are interested in identifying alternative fire suppression agents that are toxicologically and environmentally acceptable, that are as efficacious as halons in suppressing fire, and that can be deployed in existing equipment designed for halons. This latter requirement necessitates a close match between the physical properties of halon compounds and any alternative agent that will replace them. The physical properties of an agent affect the mechanisms of dispersion and thus can profoundly influence overall effectiveness. For example, halon 1301 is stored as a liquid under pressure (N_2, 600 psi) but vaporizes upon release and is dispersed as a gas. Gas-phase dispersion is a key factor in its effectiveness as a total flooding agent. Alternative agents with higher boiling points tend to be less effective overall.[9] The coexistence of liquid and gas in delivery lines complicates the dynamics of distribution, and incomplete vaporization at the nozzle can greatly impede delivery of a total flooding agent to the flame source.

This chapter describes fire suppression models and the complex chemical and physical processes involved in understanding fires and their extinguishment, outlines the chemistry of manufacture of alternative agents currently proposed by industrial suppliers, and indicates the relative performance of commercially available halon replacements. In addition, it discusses the importance of toxicological considerations in evaluating chemical candidates for replacing halon and gives a brief overview of studies conducted to identify candidate halon alternatives and further elucidate mechanisms. The chapter closes with the committee's findings regarding (1) the status of activities directed toward finding alternatives for halon as a fire suppressant and (2) the potential for discovering an environmentally acceptable equivalent.

CHEMICAL AND PHYSICAL CONSIDERATIONS IN EVALUATING ALTERNATIVE FIRE SUPPRESSION AGENTS

Several large chemical manufacturing concerns have invested heavily in the search for economically viable halon replacements, although the compounds investigated tend to be related to existing commercial products or their precursors. The National Institute of Standards and Technology (NIST) survey of candidate agents encompasses a somewhat broader range of compounds,[10] including compounds whose study will help reveal the chemistry and physics of extinguishment.

Fire Suppression Models—Chemical and Physical Mechanisms

There are a number of ways in which a fire can be extinguished. The simplest introductory concept for understanding the behavior of a fire is the fire triangle, which has three sides: fuel, oxygen, and heat. If any leg of the triangle can be removed, the fire can be extinguished. Fires are categorized as being either flaming combustion or smoldering. The former is predominantly a gas-phase phenomenon and is characterized by the emission of visible and infrared (heat) radiation. The latter type of fire occurs when solids such as plastics burn, and heterogeneous reactions at or near the surface are important. Flaming combustion, the primary target of halon 1301, generates more heat and consequently is more dangerous in the short term. Smoldering often generates more toxic gas emissions and can be more difficult to extinguish.

A fire can be extinguished by either physical or chemical mechanisms. Physical suppression mechanisms involve removing at least one leg of the fire triangle. (1) By smothering or blanketing the fire, the fuel and air are separated. An example of such a method is the use of foam extinguishers. (2) If the heat source is removed, the fire can also be suppressed. Thus, methods that cool the flame are important extinguishing techniques. For example, an agent with a high heat capacity can cool the flame by absorbing heat or can undergo a phase change that also requires heat. The most important parameters are the heat capacity and/or the latent heat of vaporization; experiments have shown that the thermal conductivity (how fast heat is transferred) is of lower importance. (3) Mechanical means such as forcing a gas over the flame at high velocity can extinguish a fire by separating the fuel and the air or the fuel and the heat. (4) For liquid or solid fuels, it is possible to place an agent that absorbs thermal radiation between the surface of the fuel and the flame. This prevents the generation of gaseous fuel and is known as flame radiation blockage.

It is also possible to have chemical suppression of a flame. This occurs when an agent or its degradation products interfere with the chain reaction that is critical to sustaining combustion. When chain carriers in or near the reaction zone are removed, the chains are disrupted and the fire cannot sustain itself.

It is possible to combine both physical and chemical effects in an agent, and in fact many of the best suppressants operate by both mechanisms. It is difficult to separate out the physical and chemical aspects of flame suppression. For example, an agent can remove heat by undergoing an endothermic decomposition process. If the decomposition products are inert, this is considered to be a physical process. However, if the products are reactive, then such products can contribute to a chemical suppression mechanism. The most likely radical chain species to be removed by a reagent or its degradation products are atomic hydrogen and oxygen and the OH radical.

There are three types of fuels for fires: gas, liquid, and solids. Gaseous fuels are the easiest to understand as they can readily participate in chain reactions. For liquids, the fuel, in general, needs to be vaporized from the liquid surface, and so the amount of fuel is determined by the rate of vaporization and the surface area of the available liquid. However, because the flame generates heat that can affect the amount vaporized, there is a complex coupling between the flame and the fuel source. Solid fuels are the most difficult to understand. The heat from the overhead flame (flame above the surface) is critical to generating the gases needed for flaming combustion. In some cases, the solid fuel can melt, forming a liquid that can be vaporized. In others, complex surface chemistry occurs, leading to the release of gaseous fuels that can feed the flame as well as to changes on the surface such as charring, which can block the emission of radiation. As is obvious from this brief discussion, there are complex flows that must be treated together with the chemistry if one is to really understand flames and fires. In the simpler case, the flow is smooth or laminar, and the adjacent layers of the fluid do not mix by mechanical means. If, however, there is random mixing of the fluids, a turbulent flow has been generated. In general vortices will be generated over a wide range of spatial scales, leading to complex mixing behavior that is not well understood.

In order to better understand how to design new fire suppression agents, it is necessary to understand some basic concepts about combustion. Given a fuel, we can write down its reaction with oxygen and calculate how much energy is released (the heat of combustion), providing we know the heats of formation of the reactants and products. For a hydrocarbon, the reaction is usually written as

$$\text{Fuel} + O_2 \rightarrow H_2O + CO_2 \,. \qquad (1)$$

Given the heat of combustion and the composition of the reaction mixture, it is possible to calculate the maximum or adiabatic flame temperature. The assumption that goes into this calculation is that all of the heat goes into heating the product gases and any other gases that are present. This calculation requires knowing the heat capacities of all of the gases. The heats of formation of all of the species are needed as well as the heat capacities in order to calculate the temperature dependence of the heat of reaction. From the expression $\Delta G = -RT\ln K$ where K is the desired equilibrium constant and ΔG is the free energy, one needs to know the free energy of the process. From $\Delta G = \Delta H - T\Delta S$, it is clear that one needs not only the enthalpy of the process (ΔH) but also the entropy (ΔS) of the process. Again, the entropies are needed for all important species. The overall equilibrium composition can then be calculated by an iterative procedure. Tables of thermodynamic properties exist, and methods have been developed for computing missing information.[11,12,13,14]

The reaction thermodynamics described above describe what happens at equilibrium but do not predict how fast the system will reach equilibrium. In order to determine the speed of the process, kinetic information is required. From the global reaction mechanism, we need to write down a reaction mechanism based on individual reaction steps, each of which is a fundamental chemical process, a unimolecular, bimolecular, or termolecular reaction. Then we have to determine the kinetics of each fundamental reaction step and use the rate constants to solve for a global kinetic rate. This is a complex process because much of the required data is not known. However, if the data are not available, methods exist for estimation.[15]

The types of reactions important in the combustion process are based on radicals.[16] If the chain reactions have long chain lengths, then the flame can continue to exist. It is important to note that combustion temperatures tend to be high, >800 K, so that generation of radicals is more important than loss due to recombination. The first reactions in the chain are the initiation reactions that lead to the initial formation of radicals. Most initiation reactions involve breaking a chemical bond and thus have high activation energies. They tend to be slow even at flame temperatures. The chain propagation reactions consume fuel or oxidizer but do not change the number of radicals; hence the chain continues. The chain may branch, leading to an increased number of radicals and hence a higher global reaction rate. These reactions generally consume the oxidizer, in most cases O_2. As most of the fuels are molecules containing carbon and hydrogen, H is an important radical in the chain. The most important branching reaction is

$$H + O_2 \rightarrow OH + O \,, \qquad (2)$$

which has an activation energy of 17 kcal/mol (due for the most part to the fact that it is an endothermic process). Note that two radicals are produced for each radical consumed. Reactions of O atoms with fuel are important, as two radicals are often produced for each O consumed. Termination reactions are those that reduce the number of active radicals or lead to formation of less reactive radicals. Many of these reactions are termolecular in nature, involving an unreactive third body such as N_2, the predominant gas in the atmosphere, and they increase in rate with decreasing temperature.

By understanding the reaction mechanism, it should be possible to design a reagent that can interfere with the production of an important radical and hence reduce the chain length. For example, reaction (2) has a competing reaction (3), which produces a less reactive radical where M is a third body,

$$H + O_2 + M \rightarrow HO_2 + M . \tag{3}$$

At 1000 K, these reactions have comparable rates, and at lower temperatures, the three-body process dominates. If H atoms are removed and the temperature can be lowered, the chains will be shorter.

The effects of thermochemistry and kinetics can be combined as the concept of thermal balance. If the flame is to continue, the energy released by the combustion process must keep the temperature high enough so that branching reactions dominate over termination reactions.

With the advent of modem high-performance computers and advanced software, it is now possible to model some of the characteristics of flames that are extremely difficult to measure.[17] Thus computational models have been developed to describe both the flame and the suppressant. The first types of models are those that describe how the agent is released. The agent must be vaporized and dispersed quickly throughout the region where the fire is occurring. The models that have been developed include the design of the holding vessel for the suppressant and most importantly the design of the exit orifice and the flow across it. This allows one to also include the use of pressurizing agents such as N_2. One can then measure the agent discharge to validate the model. Furthermore, by using modem computational fluid dynamics codes, it is also possible to use models to look at how the gas flows over a complicated interior space such as that in an engine room. Work has also been done to model simple flames, including transport and important chemical reactions. An important advance has been to use chemical reactor models such as plug flow and continuous stirred tank reactor models to predict the efficiency of suppressants. These provide good initial estimates of the chemical efficacy of a suppressant. The whole area of the modeling of reactive chemical transport is growing, and it should be possible within a few years to perform quite accurate simulations of simple flames and suppressants, given the required input data. However, it is still quite difficult to model turbulent flows and complex kinetics, and such studies will require at least teraflop computers and the design of new computational methods for treating turbulence. Studies of this type are important so that the significant investments being made in other areas of research can be leveraged to provide the information needed to design new suppressants. The models can also be used to predict concentrations of harmful by-products such as hydrofluoric acid.

As noted above, one needs to have a wide variety of thermodynamic data for designing new suppressants, especially if one is looking for chain inhibitors. It is time consuming and expensive to measure the needed thermodynamic properties if they are not known, especially for radicals. As one looks at a new agent whose thermochemistry is not known, one needs to either estimate or calculate its properties.[18] Estimation methods are showing dramatic improvements as the available databases expand. However, it is often easier to calculate thermochemical properties from ab initio quantum chemical methods than it is to measure them. Given a molecular structure, one can calculate absolute heats of formation either rigorously, with simple empirical corrections, or from idealized reactions based on known thermochemical properties. The accuracy of the results can usually be fine-tuned depending on the cost of the calculation, but it is often possible to predict heats of formation to within ±2 kcal/mol. It is just as straightforward to do this for radicals as for stable species. It is even easier to calculate the heat capacities (C_p) and entropies, as the geometries and vibrational spectra of the species need to be calculated to predict heats of formation. Statistical mechanics can then be used to provide C_p and ΔS as a function of temperature. Thus the global thermochemistry for a given mechanism can be predicted.[19]

Kinetic measurements can be made accurately, but these can be quite difficult and time consuming. Furthermore, it is often difficult to measure rates involving complex radicals or reactions that are very slow. It is also possible to calculate the rate constants from ab initio quantum chemistry and kinetic theories such as transition state theory or unimolecular rate theories. Such calculations are more difficult than the thermochemical ones, but the calculated results can often be scaled to known experimental results to extend the measured data. It is still difficult to predict rate constants to better than a factor of about five, although for simple cases much higher accuracy has been obtained.[20,21]

Halons suppress fires by both physical and chemical mechanisms. With halons, the bromine atoms are critical to disrupting the radical chains. Other agents that release halogens can have the same effect. Chlorine radicals are less effective than bromine at flame suppression, whereas iodine radicals are more effective. Unfortunately, the same halogen atoms (bromine, chlorine, and iodine) that contribute to the effectiveness of halogenated compounds, including halons, as fire suppression agents also contribute to their ozone-depleting potential. Alternative agents that do not contain halogens other than fluorine tend to be physical action agents only and therefore are generally less effective than halons.

Alternative agents that contain iodine tend to be readily tropodegradable; that is, they react in the lower atmosphere through photolysis or some other mechanism to produce stable compounds that will not deliver iodine radicals to the stratosphere. These compounds, while retaining the effective chemical mechanism of fire suppression common to halons, tend to be either unstable with respect to storage and material compatibility, or toxic, or both.

There is prior work for some suppressants in terms of their fire extinguishing mechanisms. These include the perfluorocarbons FC-116 (C_2F_6), FC-218 (C_3F_8), and FC-3-1 (C_4F_{10}); the hydrofluorocarbons HFC-125 (CHF_2CF_3), HFC-227 (CF_3CHFCF_3), HFC-318 ($CF_3CHFCHFCF_3$), HFC-3110, HFC-32/125, HFC-134a (CH_2FCF_3), and HFC-236fa ($CF_3CH_2CF_3$); the hydrochlorofluorocarbons HCFC-22 (CHF_2Cl) and HCFC-124 ($CHClFCF_3$), and CF_3I. None of these is as good a suppressant as halon 1301 in all respects. Measured or calculated data on C_1 and C_2 HFCs has been used in extensive predictions about the behavior of such compounds as suppressants. Compounds predicted from the modeling search to be possible candidates for replacing halons include CF_2HCl (HCFC-22), $CF_2{=}CClF$, $CF_2{=}CFCF_3$, $CF_2{=}CFBr$, $CF_2{=}CHBr$, CF_3I, SiF_4, and NF_3.

As discussed above, fire suppression agents act through both physical and chemical mechanisms. Physical action can extinguish flames by dilution and by providing a heat sink. Chemical action follows from interaction of the agent with the chemistry of the flame. Huggett[22] found empirically that liquid fuels would not bum if the heat capacity of the mixture exceeded about 50 cal/deg per mole O_2. Thus, a crude measure of the concentration of agent necessary to extinguish a flame by physical effects alone can be estimated by calculating the concentration that will raise the heat capacity of the surrounding atmosphere to this level. Table 2.1[23] gives the result of this calculation for several agents of interest. The difference between the observed extinguishment concentration and the calculated concentration is interpreted as a measure of the chemical effect.

Sheinson and coworkers[24] have carried through a more sophisticated analysis along these lines and have been able to measure the physical and chemical contributions to fire suppression for various agents. For example, for CF_3Br the effect is 20% physical, 25% chemical owing to CF_3, and 55% chemical owing to Br. For CF_3Cl, a less effective agent, the effect is 40% physical, 50% chemical owing to CF_3, and 10% chemical owing to Cl. CF_4 has a physical effect comparable to that of the other substituted methanes, but no chemical effect owing to the strength of the CF bond.

Table 2.1 Concentration of Agent Needed for Physical Extinguishment

Halocarbon Number	Chemical Formula	Concentration for Extinguishment		Difference
		(calculated)	(observed)	(calc. - obs.)
H-1301	CF_3Br	17.6	2.9	14.7
IFC-13I1	CF_3I	—	3.0	—
HCF-23	CF_3H	23.0	12.6	10.4
HCFC-22	CF_2HCl	21.3	11.6	9.7
HCFC-124	CF_3-CHClF	13.5	6.7	6.8
PFC-14	CF_4	20.0	13.8	6.2
HFC-236fa	CF_3-CH_2-CF_3	11.1	5.6	5.5
HCFC-123	CF_3CHCl_2	12.9	7.5	5.4
PFC-116	CF_3-CF_3	12.5	7.8	4.7
HFC-125	CF_3-CHF_2	13.8	9.4	4.4
HFC-227ea	CF_3-CFH-CF_3	10.3	6.3	4.0
PFC-218	CF_3-CF_2-CF_3	9.7	6.1	3.6
HFC-134a	CF_3-CFH_2	14.1	10.5	3.6
PFC-3-1-10	CF_3-CF_2-CF_2-CF_3	7.4	5.0	2.4

Manufacturability of Commercially Available Halocarbon Fire Suppression Agents

The manufacture of halon replacements is typically more complex than the manufacture of halons themselves. The chemical attribute that generally renders halon-like replacements easily degradable in the lower atmosphere, the hydrogen atom substituent, complicates the manufacturing process because it enables parasitic processes such as formation of significant by-products or the inactivation of catalysts used in production. Synthesis of halocarbon replacement agents often requires the use of toxic, corrosive, or otherwise difficult-to-handle precursor materials as well as severe reactor conditions. Nonetheless, economies of scale allow many of these compounds to be offered at reasonable prices, and several agents are already commercially available in quantity. Other proposed alternatives are available only in small quantities and at significantly higher prices and are thus classified as specialty chemicals. If one of these agents were to be widely adopted, it is possible that increased demand would stimulate production and lead to greater availability at lower cost.

Tables 2.2 and 2.3 list commercially available halocarbon replacements for halons in total flooding and streaming applications, respectively. The toxicological and environmental properties of the listed compounds have been well characterized.

Table 2.2 Commercially Available Replacement Agents for Total Flooding Applications

Halocarbon Number	Chemical Formula	Supplier Designation
PFC-218	CF_3-CF_2-CF_3	3M CEA-308™
PFC-3-1-10	CF_3-CF_2-CF_2-CF_3	3M CEA-410™
HFC-23	CF_3H	DuPont FE-13™
HCFC-125	CF_3-CHF_2	DuPont FE-25™
HFC-227ea	CF_3-CFH-CF_3	Great Lakes Chemical FM-200™
HFC-236fa	CF_3-CH_2-CF_3	DuPont FE-36™
HCFC-124	CF_3-CHClF	DuPont FE-241™
HCFC Blend A, a blend of:		North American Fire Guardian NAF-S-II
HCFC-22	CF_2HCl	
HCFC-123	CF_3CHCl_2	
HCFC-124	CF_3CHClF	
IFC-13I1	CF_3I	Pacific Science Triode™

Table 2.3 Commercially Available Replacement Agents for Streaming Applications

Halocarbon Number	Chemical Formula	Supplier Designation
PFC-5-1-14	CF_3-$(CF_2)_4$- CF_3	3M CEA-614™
HFC-227ea	CF_3-CFH-CF_3	Great Lakes Chemical FM-200™
HFC-236fa	CF_3-CH_2-CF_3	DuPont FE-36™
HCFC-123	CF_3CHCl_2	DuPont FE-232™
HCFC-124	CF_3CHClF	DuPont FE-241™
HCFC Blend B, which is primarily HCFC-123 (see above)		American Pacific Halotron™
HCFC Blend C, a blend of:		North American Fire Guardian NAF-P-III™
HCFC-134a	CF_3CH_2F	
HCFC-123	CF_3CHCl_2	
HCFC-124	CF_3CHClF	
IFC-13I1	CF_3I	Pacific Science Triode™

None of the agents listed in Tables 2.2 and 2.3 contain bromine, which is the chemically active species in halons. However, there is some evidence to suggest that the agents listed do effect some degree of chemical suppression. This could be owing to the production of CF_3 radicals, which may serve to chemically inhibit combustion.[25]

Synthesis of Alternative Fire Suppression Agents

This section describes the known routes of commercial manufacture of the agents of interest as replacements for halon. The synthesis of these agents requires the safe handling of exceedingly hazardous and toxic materials and intermediates. High pressures and temperatures are also required for synthesis, thus adding to the hazard. Exotic alloys must be used in the contraction of reactor vessels, because more common and less expensive materials, such as stainless steel, are subject to unacceptably high corrosion rates.

HCFC-22 (CF_2HCl) and HFC-23 (CF_3H) are both produced from chloroform by a stepwise halogen exchange reaction. This process is typically carried out in the liquid phase utilizing a Lewis acid catalyst and hydrogen fluoride as the fluorine source.[26]

$$CHCl_3 + HF \rightarrow CHFCl_2 + HCl \quad (1)$$

$$CHFCl_2 + HF \rightarrow CHF_2Cl + HCl \quad (2)$$

$$CHF_2Cl + HF \rightarrow CHF_3 + HCl \quad (3)$$

HCFC-123 (CF_3CHCl_2) can be produced commercially in at least three ways. Perchloroethylene is reacted with excess hydrogen fluoride in the presence of a Lewis acid catalyst to produce HCFC-123 directly,[27] equation (4). Alternatively, CFC-113a (CF_3CCl_3) can be hydrodechlorinated in the presence of hydrogen and a noble-metal catalyst,[28] equation (5). The third choice is to selectively chlorinate HCFC-133a (CF_3CH_2Cl) with elemental chlorine in the presence of a metal salt catalyst,[29] equation (6).

$$CCl_2{=}CCl_2 + 3HF \text{ (catalyst)} \rightarrow CF_3CHCl_2 + 2HCl \quad (4)$$

$$CF_3CCl_3 + H_2 \text{ (catalyst)} \rightarrow CF_3CHCl_2 + HCl \quad (5)$$

$$CF_3CH_2Cl + Cl_2 \text{ (catalyst)} \rightarrow CF_3CHCl_2 + HCl \quad (6)$$

HCFC-124 (CF_3CHClF) and HFC-125 (CF_3CHF_2) can be co-produced using HCFC-123 (CF_3CHCl_2) as the starting material.[30] In this vapor-phase process, excess hydrogen fluoride and HCFC-123 are passed over a transition metal halide or oxide catalyst at elevated temperatures, equation (7). If HCFC-124 is not desired, HCFC-125 can be made the sole product by increasing the temperature above 300°C and increasing the HF to HCFC-123 mole ratio in the feed stream,[31] equation (8).

$$CF_3CHCl_2 + \text{xs } HF \text{ (catalyst)} \rightarrow CF_3CHFCl + HCl \quad (7)$$

$$CF_3CHCl_2 + \text{xs } HF \text{ (catalyst)} \rightarrow CF_3CHF_2 + 2HCl \quad (8)$$

HFC-227ea (CF_3CFHCF_3) is produced in a three-step process beginning with HCFC-22 (CF_2HCl).[32] In the first step, HCFC-22 is pyrolized to produce tetrafluoroethylene and hydrogen chloride, equation (9). The tetrafluoroethylene is then pyrolized to hexafluoropropylene,[33] equation (10). Finally, hydrogen fluoride is added to yield HFC-227ea,[34] equation (11).

$$CF_2HCl + CF_2HCl \text{ (heat)} \rightarrow CF_2{=}CF_2 + 2HCl \quad (9)$$

$$CF_2{=}CF_2 + CF_2{=}CF_2 \text{ (heat)} \rightarrow CF_2{=}CFCF_3 \quad (10)$$

$$CF_2{=}CFCF_3 + HF \text{ (catalyst)} \rightarrow CF_3CFHCF_3 \quad (11)$$

HFC-236fa ($CF_3CH_2CF_3$) can be produced by at least three different processes. The first process, described by equations (12) and (13), involves two steps.[35] In the first step, perchloropropylene is reacted with hydrogen fluoride to produce CFC-216aa ($CF_3CCl_2CF_3$) and HCFC-226da ($CF_3CHClCF_3$). These intermediates are then hydrodechlorinated to produce HFC-236fa in the second step.

$$CCl_2{=}CClCCl_3 \quad + \quad HF \quad \rightarrow \quad CF_3CCl_2CF_3 \quad + \quad CF_3CHClCF_3 \tag{12}$$

$$CF_3CCl_2CF_3 \text{ and/or } CF_3CHClCF_3 \quad + \quad H_2 \quad \rightarrow \quad CF_3CH_2CF_3 \tag{13}$$

The second process[36] also involves two steps, equations (14) and (15). In the first step, vinylene chloride is condensed with carbon tetrachloride to yield HCC-230fa ($CCl_3CH_2CCl_3$). HCC-230fa is then treated with hydrogen fluoride in the liquid phase in the presence of a Lewis acid catalyst to give HFC-236fa. Alternatively, this step can be performed in the vapor phase.[37]

$$CCl_2{=}CCl_2 \quad + \quad CCl_4 \quad \rightarrow \quad CCl_3CH_2CCl_3 \tag{14}$$

$$CCl_3CH_2CCl_3 \quad + \quad HF \quad \rightarrow \quad CF_3CH_2CF_3 \tag{15}$$

HFC-236fa can also be produced from perfluoroisobutylene methyl ether, equation (16).

$$(CF_3)_2C{=}CF\text{-}O\text{-}CH_3 \quad \rightarrow \quad CF_3CH_2CF_3 + CO_2 + CH_3OH \tag{16}$$

Perfluorocarbons are commercially produced by two different process technologies. In the first process, equation (17), a hydrocarbon is passed over a hot, agitated bed of cobalt trifluoride where it is convened to the desired perfluorocarbon. The cobalt trifluoride is convened into cobalt difluoride during the process and is regenerated in a separate treatment with elemental fluorine.[38]

$$C_nH_{2n+2} \quad + \quad CoF_3(350°C) \quad \rightarrow \quad C_nF_{2n+2} \tag{17}$$

The second perfluorocarbon process involves electrochemical fluorination, i.e., the electrolysis of hydrocarbons in anhydrous hydrogen fluoride, known as the Simmons process.[39] Nickel anodes and nickel or steel cathodes are used. This method is limited to starting materials that have appreciable solubility in hydrogen fluoride. For volatile materials with little solubility in hydrogen fluoride, a complementary method that uses porous carbon anodes and a $(KF)_2HF$ electrolyte has been developed that is known as the Phillips process.[40]

IFC-13I1 (CF_3I) can be conveniently produced on the laboratory scale[41] by heating the silver salt of trifluoroacetic acid in the presence of elemental iodine. In principle, heating trifluoroacetyl chloride in the presence of potassium iodide,[42] equation (18), represents an industrially viable route to obtaining IFC-13I1.

$$CF_3COCl \quad + \quad KI \quad (\text{heat}) \rightarrow \quad CF_3I \; + \; KCl \; + \; CO \tag{18}$$

In all of the chemistry described above, many of the materials are toxic or corrosive or both. Handling and reaction on a commercial scale are therefore generally expensive. The manufacture of these materials would not be undertaken without the promise of financial reward.

Table 2.4 Relative Performance of Commercially Available Halon-Like Replacement Agents Compared to Halon 1301

CFC Number	Chemical Formula	Extinguishment (volume % required)
H-1301	CF_3Br	3.5
IFC-13I1	CF_3I	3.0
PFC-3-1-10	CF_3-CF_2-CF_2-CF_3	5.0
HFC-236fa	CF_3-CH_2-CF_3	5.4
PFC-218	CF_3-CF_2- CF_3	5.6
HFC-227ea	CF_3-CFH-CF_3	6.3
HCFC-124	CF_3-CHClF	7.0
HCFC-123	CF_3CHCl_2	7.5
PFC-116	CF_3-CF_3	7.7
HCFC-123 a	CF_3-$CHCl_2$	8.1
HFC-125	CF_3-CHF_2	9.1
HFC-236ea	CF_3-CH_2-CF_3	10.2
HFC-134a	CF_3-CFH_2	10.5
HCFC-22	CF_2-HCl	12.7
HCF-23	CF_3H	13.0
PFC-14	CF_4	18.2

SOURCE: The DuPont Company.

Effectiveness of Alternative Commercially Available Fire Suppression Agents

The various agents discussed above have been evaluated using a standard cup burner test with n-heptane fuel.[43] Table 2.4 shows the results for those agents and others, ranked in order of performance, where performance is based on the volume fraction required to extinguish the flame—the most effective agent is that which requires the lowest volume for extinguishment. Halon 1301 is included for comparison. Agents are also compared on a per mole and per weight basis normalized to halon 1301.

TOXICOLOGY ISSUES

Toxicology is a key aspect of the effort to identify suitable halon replacements. Although the Navy evacuates personnel from spaces to be flooded with halon 1301, there is always the possibility of accidental discharge, and the hazard must be carefully assessed. The toxicology of halon alternatives has been studied, and protocols for agent testing and use have been delineated.

In evaluating the toxicology of chemical candidates for replacing halons as fire suppressants, the key consideration is that potential exposures, although infrequent, will possibly be at high levels. The population exposed will be under stress and possibly engaging in a high level of physical activity. If effects are seen in studies focusing on brief, high-level exposures, additional follow-up studies may be desirable to further characterize and understand them. In addition, some consideration should be given to carcinogenic potential.

The initial study of a candidate replacement for halon should be an acute inhalation toxicity study. The study should be robust enough to define the dose response curve as well as the median lethal concentration. Typically, these studies involve single exposures to groups of five male and five female rats for a period of 4 hours. Mice can also be used and can be the species of choice when the substance being tested is in short supply. Although 4 hours is a long period of time relative to the brief exposure (typically 1 to 2 minutes) expected under conditions of actual use, this time period has become a comparative standard and allows for selection of substances based on a common protocol. Should lethality be a concern, an additional study for a shorter time period could subsequently be conducted. Recently, more attention has been given to predicting the effects of brief exposures to suppressants, and protocols have been developed that include exposure periods of 1 hour, 30 minutes, and even 10 minutes. Information on exposure for brief periods is being used by the U.S. EPA-sponsored Federal Advisory Committee on Acute Exposure Guidance levels,[44] and more information on the effects of brief exposures will likely be available from future studies.

In addition to observations for lethality, detailed clinical observations of test animals should be conducted during and after exposure. It is common to observe animals at hourly intervals during and immediately after exposure, and then daily for the remainder of a study, to allow determination of other effects such as central nervous system depression (many of the chemicals used as fire suppressants are anesthetics); neurotoxic effects, such as development of convulsions or impaired movement; or irritation of the eyes or respiratory system. Surviving test animals are held for 2 weeks for observation of postexposure effects and recovery. Body weights are measured frequently and can be sensitive indicators of an adverse response to exposure. Some key organs, such as the lungs, kidney, liver, testes, and heart, may be weighed and examined microscopically for abnormal changes, although such examination rarely produces useful information because test animals tend to recover during the 14-day observation period.

If only minimal effects of exposure are seen in test animals, the next phase of testing can be initiated. If the compound being tested shows marked toxicity, it may be eliminated from consideration as a flame suppressant in occupied areas. If the substance exhibits intermediate toxicity, additional testing may be warranted. If the chemical at relatively low levels induces sleep (central nervous system depression), it may be desirable to evaluate it for anesthetic potential to determine if its use could create a hazard by reducing the alertness of people exposed to it. If signs of neurotoxicity, such as impaired movement or convulsions, are seen, a study designed to look at those endpoints may be indicated. A basic protocol has been developed by the Organisation for Economic Cooperation and Development [45] and can be used as a reference point. In this application it may be desirable to add some of the parameters described above.

Information on a candidate substance's potential to cause skin[46] and eye irritation[47] may also be valuable for use in protecting maintenance personnel and those other individuals going into an area after discharge of a fire suppressant and coming into contact with walls, furniture, and other items that have been contaminated.

Many of the chemicals used as flame suppressants have the potential to sensitize the heart to the effects of adrenaline. A protocol for evaluating this endpoint has been developed[48] and was reviewed recently by the National Research Council's Committee on Toxicology. This is a sensitive test because administration of adrenaline by injection increases blood levels well above those seen under conditions of stress.[49,*] If a response is seen at a concentration lower than one that could be encountered following discharge of a flame suppressant, this effect should be carefully considered. While there has been extensive discussion about the most appropriate method for extrapolation of this data for use in human

[*] Briefly, in this study, dogs were exposed to the test compound at a set exposure level for a period of 5 minutes, during which time the blood level of the chemical approached near-equilibration levels. The test dogs were then given an injection of adrenaline at a level just below that necessary to cause a cardiac arrhythmia (a rapid, irregular, potentially fatal heart beat). Each dog's EKG pattern was continuously measured during this procedure. The appearance of a response to the adrenaline injection was taken as a positive response.

risk assessment, in fact, estimates of exposures associated with the development of cardiac arrhythmias in humans have always been much higher than the threshold seen in dogs. This has led to a direct application of the no-effect levels observed in these dog studies to human risk assessments. In the example above with halon 1301, the recently reported no-effect levels are 75,000 to 100,000 ppm in two dog studies. Comparing this to the projected levels for flooding at 50,000 to 70,000 ppm indicates that the application levels are appropriate for use in an occupied area.

It is desirable to conduct a multiple-exposure study to determine what effects may result from repeated low-level exposures to fire extinguishing agents, such as are encountered in the workplace or during maintenance operations, especially if indications of toxicity were seen at exposure levels approaching those that might be encountered under actual discharge or use conditions. Either a 4-week[50] or 13-week[51] exposure period would be appropriate. The 13-week design may be more desirable if frequent exposures are anticipated. Such a study normally involves a series of daily 6-hour exposures for either 5 or 7 days per week using four groups of 10 male and 10 female rats in each group. Recovery groups of five animals per sex may be included. These animals are held for 2 weeks following the final exposure to assess the animal's potential to recover from the effects of exposure. During the study, in addition to measurements of body weight and clinical observations, complete blood counts and serum chemistry analysis are conducted. At the end, the animals are sacrificed and all major organs are examined for effects of the exposure. If a recovery group has been included, the same parameters are also evaluated for this group.

When there is the possibility of pregnant women being exposed repeatedly to a chemical agent for fire suppression, it would be desirable to conduct a developmental toxicity study.[52] Information on the potential of a chemical agent to cause either birth defects or developmental toxicity can be obtained from an inhalation developmental toxicity study. In this type of study, pregnant animals, usually rats, mice, or rabbits, are exposed to the chemical during the organ-development phase of pregnancy. At the end of the pregnancy the offspring are evaluated for abnormalities that could have been caused by the mother's exposure to the chemical. The most critical findings would be frank birth defects; however, other effects such as birth weight are also evaluated. A single species, either the rat or mouse, should provide adequate information. In the event of a positive finding, a second study, with rabbits, should be considered. The results from these studies, along with an estimate of potential exposure levels, should be considered in determining the potential risk to pregnant women exposed to chemical fire suppressants during maintenance or other procedures and in area discharges.

Mutagenicity is the last element that should be considered in evaluating the toxicology of candidate replacements for halon. Although a carcinogenic study would not be a part of an evaluation program, conducting a few studies to evaluate mutagenic potential would aid in our understanding of carcinogenic potential. Two that are particularly useful are the Ames assay and the human lymphocyte chromosome aberration assay. Both are run with cell cultures. In the Ames assay, bacteria are exposed to a test chemical and their potential to revert to a form capable of growing in a medium lacking a normally essential nutrient is evaluated. For this change to occur, the bacteria must mutate. In the second assay, the direct effect of exposure to a chemical on the chromosomes in a human cell line is evaluated using a microscope. Breaks in the chromosomes are positive indicators. Other genetic studies, including those involving whole animals, such as the micronucleus assay, may be considered for special programs.

While other specialized studies involving metabolism, uptake, and pharmacokinetics may be indicated from time to time based on preliminary results, their selection would be compound specific. These studies are usually conducted when the levels of concern from one or more of the toxicology studies are near the potential human exposure level. They answer questions such as whether a test animal metabolizes a chemical in the same way as a human does, and whether the dose of a chemical actually reaching an organ system of concern would be higher or lower in a human compared with a test animal exposed at the same airborne level. The results from such studies allow for a more precise assessment of risk for humans exposed to particular chemical compounds. Generally, a program looking at acute inhalation toxicity, cardiac sensitization, potential developmental toxicity, and possibly the effects of repeat exposure, along with a limited genetics screen, should be adequate to characterize the toxicity of potential substitute flame suppressant materials. Table 2.5 lists several results of such toxicology studies for some of the compounds of interest.

Table 2.5 Summary of Selected Toxicity Data for Possible Halon Substitutes

	4-hr LC_{50} (ppm)		Developmental Toxicity	Genetic Assays	
Substitute				Ames	Chromosome Aberration
HCFC-22[a]	>200,000	50,000	Possible slight effect rat—50,000 No effect rabbit—50,000	Not active	Not active
HCFC-123[b]	32,000	20,000	No adverse effects rat and rabbit	Not active	Positive (not active, 5 other genetic assays)
HCFC-124[c]	>360,000	25,000	No adverse effects rat and rabbit	Not active	Not active
HFC-125[d]	>800,000	75,000	No adverse effects rat and rabbit	Not active	Not active
HFC-134a[e]	>500,000	80,000	No adverse effects rat and rabbit	Not active	Not active
HFC-227ea[f]	>800,000	100,000	No adverse effects rat and rabbit	Not active	Not active
HFC-236fa[g]	>457,000	150,000	No adverse effects rat and rabbit	Not active	Not active

[a] HCFC-22. Joint Assessment of Commodity Chemicals No. 9 Chlorodifluoromethane. European Centre for Ecotoxicology and Toxicology of Chemicals, Brussels, Belgium (October, 1989).
[b] HCFC-123. Joint Assessment of Commodity Chemicals No. 33 1, 1-Dichloro-2,2,2-trifluoroethane. European Centre for Ecotoxicology and Toxicology of Chemicals, Brussels, Belgium (February 1996).
[c] HCFC-124. Joint Assessment of Commodity Chemicals No. 25 1-Chloro-1,2,2,2-tetrafluoroethane. European Centre for Ecotoxicology and Toxicology of Chemicals, Brussels, Belgium (July 1994).
[d] HCFC-125. Joint Assessment of Commodity Chemicals No. 24 Pentafluoroethane. European Centre for Ecotoxicology and Toxicology of Chemicals, Brussels, Belgium (May 1994).
[e] HFC 134a. Joint Assessment of Commodity Chemicals No. 31 1, 1,2-Tetrafluoroethane. European Centre for Ecotoxicology and Toxicology of Chemicals, Brussels, Belgium (February 1995).
[f] Personal communication from Great Lakes Chemical Corp., One Great Lakes Blvd., W. Lafayette, IN 47906.
[g] Toxicologic Profile of the Alternative Refrigerant, HFC-236FA, R. Valentine, D.A. Keller, et al. *The Toxicologist*, No. 30 1, 2, Abstract 1489 (March 1996).

On exposure to flames, halons and halocarbon alternatives react to yield toxic chemicals. CO and CO_2 are present in fire atmospheres generally, but halocarbons produce HF, HCl, HBr, and COF_2. These chemicals are produced during extinguishment with halon 1301, but in much smaller quantities. The amounts of these dangerous compounds can be minimized by reducing the time to extinguish the fire, increasing the amount of agent employed, and discharging the agent rapidly into the fire volume. The importance of this source of toxicity depends on cool-down time before reentry of the affected space. Design in this connection will be agent and practice specific.[53,54]

BRIEF OVERVIEW OF STUDIES ON HALON SUBSTITUTES AND MECHANISMS

In 1986, the U.S. Air Force recognized the potential for regulation of halons and initiated a program to identify halon replacements. This program demonstrated that halon replacements with decreased or zero ODP could be obtained by (1) eliminating bromine and (2) substituting hydrogen for fluorine to decrease atmospheric lifetime.[55] The Air Force program identified HCFCs, HFCs, and PFCs as the most likely near-term replacements.

The National Institute of Standards and Technology (NIST) issued a report in 1990 whose objective was to initiate a systematic search for halon replacements.[56] This report is firmly grounded in the science and technology of combustion and current understanding of the role of halogenated compounds in ozone depletion in the stratosphere. The centerpiece of the publication is a list of 103 chemicals "covering a range of chemical and physical principles thought to affect flame suppression capability." The list includes saturated halocarbons, halogenated ketones, esters and anhydrides, unsaturated halocarbons, halogenated ethers, sulfur halides, compounds containing phosphorus, silicon, germanium and metals, and inert gases. The list contains both candidate halon replacement agents and compounds chosen to test principles of fire suppression. While the NIST list is not claimed to be exhaustively inclusive, it is sufficiently extensive to suggest that the most obvious classes of alternative agents are included. The list includes most of the commercially available agents listed in Tables 2.2 and 2.3, omitting HFC-227ea and HFC-236fa. These compounds were later considered by NIST for aircraft nacelle and dry bay applications.[57,58]

Flame suppressants that do not endanger the ozone layer either do not contain the heavy halogens (chlorine, bromine, or iodine) or have atmospheric lives so short that they do not reach the stratosphere. Perfluorocarbons satisfy the first condition but do not offer chemical flame suppression activity and are somewhat less effective in extinguishing fires. Compounds that contain the heavy halogens but are destroyed before they reach the stratosphere are of special interest as halon replacement agents. Substitution of H for a halogen, increasing the dipole moment, and shifting the absorption toward the red are strategies pursued in the NIST list, including compounds containing hydride, ketone, ester, anhydride, and double or triple bonds.

In connection with halon 1301 replacement agents, only a limited number of compounds have matching physical properties, i.e., are storable as a liquid at normal ambient temperatures, but have a high vapor pressure and so evaporate rapidly on discharge. Of the 103 compounds listed by NIST, only 31 have boiling points below 0°C (the b.p. for halon 1301 is -68°C), and some of these are toxic or have interest only for determining mechanisms. A low boiling point and high vapor pressure strongly indicate compounds of low molecular weight and/or high fluorine content. The point is that the number of candidate halocarbon halon replacement agents is not very large.

Tapscott and coworkers[59] are testing various classes of compounds as candidates for total-flooding fire suppression, including brominated fluoro ethers, alkenes and aromatics, amines, and carbonyls, as well as phosphorus and silicon compounds. Some of these materials have good fire suppression characteristics, but satisfactory performance in terms of toxicity, materials compatibility, and other properties remains to be proved.

FINDINGS

FINDING: After reviewing research, development, toxicology, and engineering activities directed toward finding alternative and replacement agents for halon 1301 and halon 1211, the committee finds that in this context, the relevant aspects of the problem are being studied effectively and a comprehensive body of scientific and engineering knowledge is being developed, and the committee has identified no obvious gaps in these important efforts.

FINDING: It is unlikely that a drop-in replacement agent will be discovered that will exhibit all of the beneficial properties of halon 1301 and not also exhibit a significant environmental impact .

REFERENCES

1. *An Appraisal of Halogenated Fire Extinguishing Systems*, National Academy of Sciences, Washington D.C. (1972).
2. *Halogenated Fire Suppressants*, R.G. Gann, Ed., American Chemical Society, Washington D.C. (1975).
3. W.M. Pitts, M.R. Nyden, R.G. Gann, W.G. Mallard, and W. Tsang, *Construction of an Exploratory List of Chemicals to Initiate the Search for Halon Alternatives*, NIST Technical Note 1279, U.S. Department of Commerce, Washington D.C. (1990).
4. *Evaluation of Alternative In-Flight Fire Suppressants for Full-Scale Testing in Simulated Aircraft Engine Nacelles and Dry Bays*, R.G. Gann and W.M. Pitts, Eds., U.S. Department of Commerce, Washington D.C. (1994). Note particularly Chapter 5: "Flame Inhibition Chemistry and the Search for Additional Fire Fighting Chemicals," pp. 467-641.
5. *Halon Replacements*, A.W. Miziolek and W. Tsang, Eds., American Chemical Society, Washington D.C. (1995).
6. *Fire Suppression System Performance of Alternative Agents in Aircraft Engine and Dry Bay Laboratory Simulations*, R.G. Gann, Ed., SP890, U.S. Department of Commerce, Washington D.C. (1995).
7. J.F. Malcolm, *NFPA Quarterly* **45**, 119 (1951).
8. M.J. Molina and F.S. Rowland, *Nature* **249**, 810 (1974).
9. *Evaluation of Alternative In-Flight Fire Suppressants for Full-Scale Testing in Simulated Aircraft Engine Nacelles and Dry Bays*, R.G. Gann and W.M. Pitts, Eds., U.S. Department of Commerce, Washington D.C. (1994), p. 305.
10. W.M. Pitts, M.R. Nyden, R.G. Gann, W.G. Mallard, and W. Tsang, *Construction of an Exploratory List of Chemicals to Initiate the Search for Halon Alternatives*, NIST Technical Note 1279, U.S. Department of Commerce, Washington D.C. (1990).
11. D.A. Dixon and K.D. Dobbs, "Advanced Materials," *Transactions of the Material Research Society of Japan* **18A** (1993); D.A. Dixon and K.D. Dobbs, Ecomaterial, V, R. Yamato et al., Eds., Elsevier, Amsterdam , (1994) p. 373.
12. D.A. Dixon, *J. Phys. Chem.* **92**, 86 (1988).
13. D.A. Dixon and B.E. Smart, *Chem. Eng. Commun.* **98**, 173 (1990).
14. F. Burgess, Jr., M.R. Zachariah, W. Tsang, and P.R. Westmoreland, "Thermochemical and Chemical Kinetic Data for Fluorinated Hydrocarbons," NIST Technical Note 1412, U.S. Department of Commerce Washington, D.C. (1995).
15. K.D. Dobbs, D.A. Dixon, and A. Koornicki, *J. Phys. Chem.* **98**, 8852 (1993).
16. Pitts et al., NIST Technical Note 1279 (1990), pp. 4-8
17. M.D. Smooke and V. Giovanagigli, *Int. J. Supercomput. Appl.* **5**, No. 4, 34 (1991).
18. E.S. Domelski and E.D. Hearing, *J. Phys. Chem. Ref. Data* **22**, 805 (1993).
19. *An Appraisal of Halogenated Fire Extinguishing Systems*, National Academy of Sciences, Washington D.C. (1972).
20. *An Appraisal of Halogenated Fire Extinguishing Systems*, National Academy of Sciences, Washington D.C. (1972).
21. *Halogenated Fire Suppressants*, R.G. Gann, Ed., American Chemical Society, Washington D.C. (1975).
22. C. Huggett, *Combustion and Flame* **20**, 140 (1973).
23. R. Fernandez, private communication.
24. R.S. Sheinson, J.E. Penner-Hall, and D. Indritz, *Fire Safety J.* **15**, 437 (1989).
25. R.S. Sheinson, J.E. Penner-Hall, and D. Indritz, *Fire Safety J.* **15**, 437 (1989).
26. Daudu and M.A. Youker, U.S. Patents 2,005,705 and 2,005,709 (1935).
27. L.E. Manzer and V.N.M. Rao, U.S. Patent 4,766,260 to du Pont (1985).
28. S. Morikawa, M. Yoshitake, and S. Tatematsu, Japanese Patent Appl. 1-319440 (1989).
29. Y. Furutaka, Y. Homota, and T. Honda, Japanese Patent Appl. 1-2900638 to Daikin (1989).
30. *B. Cheminal*, E. Lacroiz and A. Lantz, European Patent Appl. 638535 to Elf Autochem (1994).
31. P. Cuzzato, A. Masiero, and F. Rinaldi, European Patent Appl. 638535 to Ausimont (1994).
32. J.M. Hamilton, Jr., in *Advances in Fluorine Chemistry*, Vol. 3, Butterworth, London (1963).
33. D.A. Nelson, U.S. Patent 2,758,138 to du Pont (1956); E.H. Ten Eyck and G.P. Larson, U.S. Patent 2,970,176 to du Pont (1961); H. Niimiya, British Patent 1,016,016 to Daikin (1963); N.E. West, U.S. Patent 3,873,630 to du Pont (1972).
34. S.P. Von Halasz, German Patent 2,712,732 to Hoechst (1978).

35. K. Gassen et al., U.S. Patent 5,171,901 to Bayer (1992).
36. S. Morikawa, M. Yoshitake, and S. Tatematsu, Japanese Patent Appl. 1-319440 (1989).
37. M.S. Nappa and A.C. Sievert, U.S. Patent 5,414,165 to du Pont (1954).
38. R.J. Kingdom, G.D. Bond, and W.L. Linton, British Patent 1,281,822 to ISC (1970).
39. W.V. Childs et al., U.S. Patent 5,387,323 to 3M (1991).
40. R.A. Paul and M.B. Howard, U.S. Patent 3,840,445 to Phillips Petroleum (1975).
41. P. Tarrant, *J. Fluorine Chem.* **25**, 69 (1984).
42. C. Krespan, *J. Org. Chem.* **23**, 2016 (1958).
43. *NFPA Standard on Clean Agent Fire Extinguishing Systems*, P.J. DiNenno, Ed. (1996), p. 65.
44. National Advisory Committee on Acute Exposure Guidelines Levels, U.S. EPA Office of Prevention, Pesticides and Toxic Substances, 60 FR 55376, October 31 (1995).
45. *Acute Toxicity (Inhalation)*, Official Journal of the European Communities, English Edition, Legislation **35**, L383A/117, December 29 (1992).
46. *Acute Toxicity (Skin Irritation)*, Official Journal of the European Communities, English Edition, Legislation **35**, L383A, December 29 (1992).
47. *Acute Toxicity (Eye Irritation)*, Official Journal of the European Communities, English Edition, Legislation **35**, L383A, December 29 (1992).
48. C.F. Reinhardt, A. Azar, M.E. Maxfield, P.E. Smith, Jr., and L.S. Mullin, "Cardiac Arrhythmias and Aerosol 'Sniffing,'" *Archives of Environmental Health* **22**, 265, February (1971).
49. Committee on Toxicology, National Research Council, *Toxicology of Alternatives to Chlorofluorocarbons: HFC-134a and HCFC-123*, Chapter 2, pp. 13-22, National Academy Press, Washington, D.C. (1996).
50. *Repeated Dose (28 Days) Toxicity (Inhalation)*, Official Journal of the European Communities, English Edition, Legislation **35**, L383A/140, December 29 (1992).
51. Organisation for Economic Cooperation and Development (OECD). Subchronic Inhalation Toxicity: 90-day Study. Method 413. OECD Guideline for Testing of Chemicals, 12 May (1981).
52. Organisation for Economic Cooperation and Development (OECD). Teratology. Method 414. OECD Guideline for Testing of Chemicals, 12 May (1981).
53. S.R. Skaggs, T.A. Moore and R.E. Tapscott, p. 99 in *Halon Replacements* , A.W. Miziolek and W. Tsang, Eds, American Chemical Society, Washington, D.C. (1995).
54. R. Sheinson, ibid., p. 175.
55. R.E. Tapscott, et al., Five reports from Air Force Engineering and Services Laboratory, Tyndall Air Force Base, Florida (1987-1990). Also issued by New Mexico Engineering Research Institute: WA3-31(103.21), SS2.06(1), WA3-79(3.21), WA3-78(3.21), SS2.06(2).
56. W.M. Pitts, M.R. Nyden, R.G. Gann, W.G. Mallard, and W. Tsang, *Construction of an Exploratory List of Chemicals to Initiate the Search for Halon Alternatives*, NIST Technical Note 1279, U.S. Department of Commerce, Washington D.C. (1990).
57. *Evaluation of Alternative In-Flight Fire Suppressants for Full-Scale Testing in Simulated Aircraft Engine Nacelles and Dry Bays*, R.G. Gann and W.M. Pitts, Eds., U.S. Department of Commerce, Washington D.C. (1994). Note particularly Chapter 5: "Flame Inhibition Chemistry and the Search for Additional Fire Fighting Chemicals," pp. 467-641.
58. *Halon Replacements*, A.W. Miziolek and W. Tsang, Eds., American Chemical Society, Washington D.C. (1995).
59. T.A. Moore, C.A. Weitz, and R.E. Tapscott, pp. 551-567 in *Proceedings of the Halon Options Technical Working Conference*, Albuquerque, N. Mex., May 7-9 (1996).

3

Atmospheric Chemistry and Evaluation of Environmental Effects of Fire Suppressants

Laboratory studies, atmospheric measurements, and numerical models of the atmosphere have provided important evidence for the significant effects of chlorine and bromine on stratospheric ozone in the last few decades and have confirmed the role of bromine-containing halons in ozone depletion. This chapter examines compounds being considered as replacements for halon and the approaches being used to evaluate their potential effects on Earth's stratospheric ozone and climate, including their contribution to radiative forcing—the greenhouse effect—and other possible environmental impacts. The chapter's opening section briefly describes how researchers came to learn of the link between release of man-made chlorine- and bromine-containing compounds and depletion of ozone.

REDUCTION OF OZONE—MECHANISMS AND EFFECTS

Earth's atmosphere is made up of regions, classified by the vertical trends in temperature (Figure 3.1). The stratosphere, where most of the atmospheric ozone resides (see Box 3.1), extends from ~10 to 50 km above the surface. Stratospheric ozone is important because it absorbs much of the intense ultraviolet (UV) radiation from the sun, dissociating into O_2 and O. The O_2 and O recombine to form ozone, which can then absorb more radiation. When it is absorbed by cells, UV radiation can lead to cellular damage, which can be manifested directly as cancer, for example, or as DNA damage that is transferred to later generations.

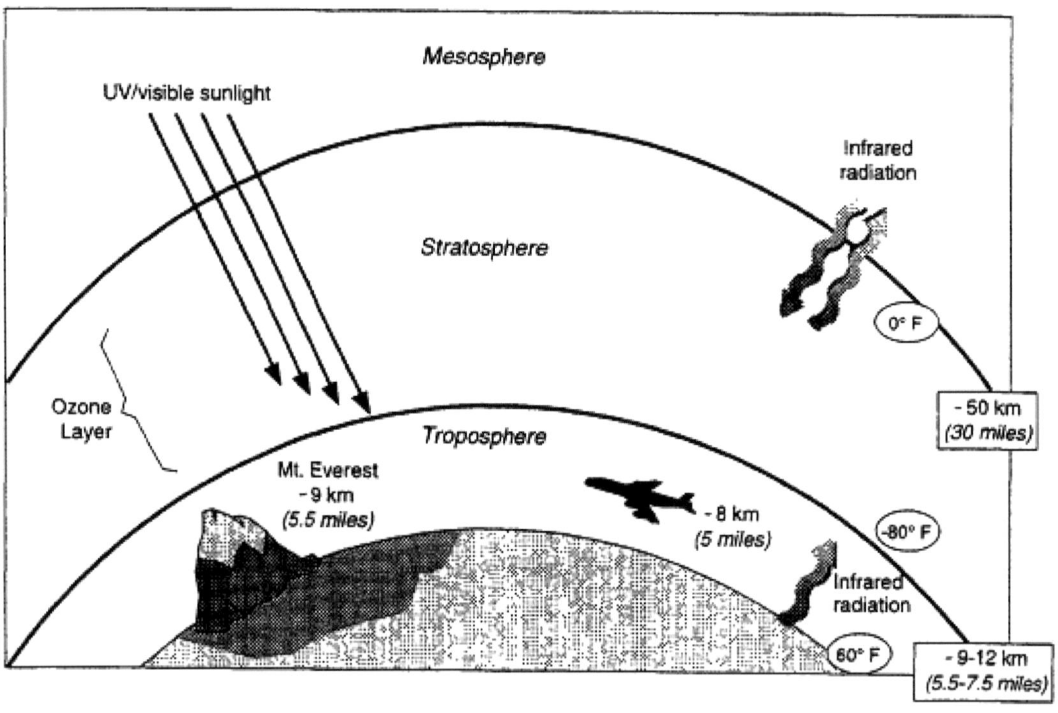

Figure 3.1
Regions of Earth's atmosphere. Roughly 90% of the ozone is in the stratosphere, with most of the rest in the troposphere.

BOX 3.1 HOW MUCH OZONE IS THERE IN THE STRATOSPHERE?

If all the ozone in the atmosphere were brought down to sea level, it would be merely 2.5 to 4.5 mm thick. However, this amount of ozone can absorb most of the harmful UV radiation entering Earth's atmosphere. Atmospheric scientists use the Dobson Unit (DU) to measure the amount of ozone overhead, i.e., from the ground to outside the atmosphere. 1 DU is equal to 1 millicentimeter of ozone at sea-level pressure (0°C). So, in general, the amount of column ozone is between 250 and 450 DU. In Antarctica, the ozone column has been measured to be as low as 125 DU, i.e., a reduction of ozone by a factor of two or three.

The potential for ozone depletion in the stratosphere became an important topic when the first commercial supersonic transport aircraft were proposed in the late 1960s.[1] At that time, scientists noted that nitrogen oxide (NO_x) emissions from engine exhaust associated with aircraft operating in the stratosphere could be involved in a catalytic cycle leading to ozone loss.[2] This realization helped lead to the U.S. withdrawal from this potential market. In 1974, Rowland and Molina[3] proposed that chlorofluorocarbons (CFCs) emitted by human activities at Earth's surface could pass though the troposphere to the stratosphere. In the stratosphere, UV radiation from the sun is absorbed by the C-Cl bonds in CFCs, leading to rupture of the bonds and liberation of Cl atoms. The Cl atoms can participate in a catalytic cycle that leads to conversion of ozone to oxygen molecules. Later laboratory studies showed that bromine and iodine atoms also participate in similar reactions and can also affect ozone.[4] The recognition that anthropogenic emissions of gas could lead to destruction of ozone in the stratosphere led to the Nobel Prize in chemistry being awarded in 1995 to P. Crutzen, M. Molina, and F.S. Rowland.

Although the amounts of CFCs and halons released into the atmosphere are small in terms of the total amount of gas there, they have a great impact on the global ozone balance for three reasons: (1) Ozone is in a constant state of "flux"—it is made and destroyed by natural processes that define a delicate balance (Box 3.2). (2) The production of ozone is controlled by solar input that does not undergo dramatic fluctuations. Removal of ozone from the atmosphere is controlled by catalytic processes, set in motion by small concentrations of natural and synthetic chemicals that can destroy a large number of ozone molecules without being destroyed in the process. Depletion is accelerated when the halogen atoms chlorine, bromine, and iodine are present; thus changes in the "balance" lead to a lower level of ozone. (3) A large fraction of the anthropogenic (and natural) reagents released at Earth's surface can be transported to the stratosphere if they are chemically stable in the troposphere. Because halons and CFCs are very stable in the troposphere, a large fraction of the released amounts reach the stratosphere, where they are quickly broken apart to release halogens that are active in destroying ozone. The parameter that defines how much of the released amounts reach the stratosphere is the atmospheric lifetime (Box 3.3).

Initially, ozone depletion was just a hypothesis based on laboratory data. It was unclear whether other trace gases in the atmosphere would interfere with the ozone removal cycles. Starting in the early 1970s, numerical models were developed to simulate the interactions of trace gases under atmospheric conditions, and the models showed that ozone depletion should occur. The trace gas concentrations simulated by these models compare favorably with measured concentrations, which lends credence to the models. However, direct comparison of model-calculated change in ozone with observed change is more difficult. Because the expected globally averaged ozone decrease is small, on the order of a few percent over a decade, detecting such change in the atmosphere requires extracting this long-term trend from large seasonal cycles (~30%) and interannual variations (~10%). It was not until the early 1980s that ground-based and satellite data were sufficient to determine a clear trend.

Direct measurements now show that stratospheric ozone depletion has occurred during the past two decades.[5] In fact, the extent of ozone depletion is *larger* than predicted by the models based on gasphase chemistry. It has been shown in laboratory experiments that heterogeneous (gas-particle) chemistry on cold particles can occur such that a Cl atom can destroy even more ozone molecules than thought previously.[6] The change in the vertical distribution of ozone has been measured and the trend deduced.

BOX 3.2 HOW IS OZONE, O_3, MADE AND DESTROYED IN THE STRATOSPHERE?

What Determines the Ozone Abundance?

Ozone is produced via the photolysis of molecular oxygen by solar radiation: $O_2 \xrightarrow{UV} O + O$. The oxygen atoms combine with O_2 to make O_3: $O + O_2 \rightarrow O_3$. Ozone itself is photolyzed to make O atoms, but most of the O atoms recombine with O_2 to make ozone. However, a small fraction of the O atoms react with ozone to destroy ozone: $O + O_3 \rightarrow 2\,O_2$. So, there is a natural balance in the stratosphere between the formation of ozone and its destruction. This chemical scheme, first proposed by S. Chapman,[1] is referred to as the Chapman mechanism.

In addition to O atoms, other reactants, many of which are naturally occurring, can also destroy ozone. For example, OH and HO_2 radicals, which are present in the stratosphere, can *catalytically* destroy ozone: $OH + OH_3 \rightarrow HO_2 + O_2$, $HO_2 + O_3 \rightarrow OH + 2\,O_2$. Notice that in this sequence of reactions, OH and HO_2 are not lost, but two molecules of ozone are destroyed. This is the concept of homogeneous gas-phase free radical catalysis and the reason that a small abundance of reactive species can destroy a large amount of ozone.

As with OH and HO_2, chlorine and bromine can also destroy ozone. Examples include: $Cl + O_3 \rightarrow ClO + O_2$, $ClO + O \rightarrow Cl + O_2$, whose net result is the destruction of one ozone molecule and an O atom, which would have ended up as an ozone molecule. The most important catalytic reactions involving bromine and chlorine together are $Br + O_3 \rightarrow BrO + O_2$, $Cl + O_3 \rightarrow ClO + O_2$, $ClO + BrO \rightarrow Cl + Br + O_2$, which leads to a net reaction: $2\,O_3 \rightarrow 3\,O_2$. Another such sequence, which involves bromine with a naturally occurring species, is $Br + O_3 \rightarrow BrO + O_2$, $BrO + HO_2 \rightarrow HOBr + O_2$, $HOBr \xrightarrow{light} OH + Br$, $OH + O_3 \rightarrow HO_2 + O_2$, which also destroys two molecules of ozone. These reactions involving bromine are especially effective in the lower stratosphere, where much of the observed ozone depletion occurs. In addition to these catalytic cycles, there are a very large number of cycles involving nitrogen oxides, hydrogenated species, and halogen species.

The balance between production, via photolysis of oxygen, and loss via catalytic ozone destruction cycles, described above, determines the mean level of ozone in the atmosphere. Since the production rate of ozone is essentially constant, any enhancements in the loss processes, such as introduction of bromine compounds into the stratosphere, will lead to a lower level of ozone.

[1] S. Chapman, "A Theory of Upper Atmospheric Ozone," *Memoirs of the Royal Meteorological Society* **3** (26), 103-125 (1930).

The ozone depletion observed in the upper stratosphere (Figure 3.2) is consistent with the Rowland and Molina hypothesis. For the lower stratosphere, where the majority of the ozone loss has occurred, it is now clear that such reactions also take place in/on sulfuric acid aerosols, which are always present at low levels in the stratosphere. Volcanic eruptions can greatly enhance the number of these droplets and increase the effectiveness of bromine and chlorine in destroying ozone. Thus, following the eruption of Mt. Pinatubo in 1991, there was a measurable decrease in stratospheric ozone abundance. The large loss of ozone during the 1990s in the lower stratosphere can be attributed, in part, to the eruption of Mt. Pinatubo, which increased the number of particles on which heterogeneous chemistry can occur and helped make the connection between the role of halogen chemistry and ozone changes.[7] A semi-quantitative understanding of the entire ozone loss has emerged. The release of man-made chlorine and bromine compounds is the primary cause of ozone depletion.

The Antarctic ozone hole, formed in the Antarctic stratosphere during the springtime, caught the attention of the atmospheric sciences community in the mid-1980s.[8] During August and September 1987—the end of winter and beginning of spring in the Southern Hemisphere—aircraft equipped with many different instruments for measuring a large number of chemical species were flown repeatedly over Antarctica.[9] Among the chemicals measured were ozone and chlorine oxide, the reactive chemical identified in the laboratory as one of the participants in ozone-destroying chain reactions. On the first flights southward from the southern tip of South America, relatively high concentrations of ozone were measured everywhere over Antarctica. By mid-September, however, the instruments recorded low

BOX 3.3 CONCENTRATIONS AND LIFETIMES

The tropospheric concentration of a halon or a replacement compound is dependent on the rate of its emission into the atmosphere and its atmospheric lifetime. Measurements of halons 1211 and 1301 show that their current global mixing ratios (the ratio of volume density or concentration to the volume density of air) are about 2.5 and 2.0 parts per trillion by volume (pptv), respectively, and are currently increasing at about 2 and 6% per year, respectively.[1,2,3] These rates of increase have slowed appreciably in recent years, consistent with the reduction in production and emission of these compounds. Despite such small concentrations, production of these compounds has been halted because of the capability of the bromine they contain to destroy ozone. Atmospheric models indicate that halons 1211 and 1301 are essentially nonreactive in the troposphere and are destroyed through photolysis in the stratosphere, resulting in atmospheric lifetimes of about 20 and 65 years, respectively.[4,5] Because of their long atmospheric lifetimes, the destruction of these halons generally releases their bromine into the stratosphere, where the bromine has the greatest impact on ozone. The long atmospheric lifetimes also imply that halons already emitted will be releasing bromine into the stratosphere for several decades to come.

[1] J.H. Butler, J.W. Elkins, B.D. Hall, S.O. Cummings, and S.A. Montzka, *J. Geophys. Res.* **359**, 403-405 (1992).
[2] World Meteorological Organization, *Scientific Assessment of Ozone Depletion*, (1994); World Meteorological Organization, Global Ozone Research and Monitoring Project, Report No. 37D, Geneva (1995).
[3] Intergovernmental Panel on Climate Change (IPCC), *Radiative Forcing of Climate Change*, Cambridge University Press, Cambridge, Great Britain (1996).
[4] World Meteorological Organization, *Scientific Assessment of Ozone Depletion*, (1994); World Meteorological Organization, Global Ozone Research and Monitoring Project, Report No. 37D, Geneva (1995).
[5] Intergovernmental Panel on Climate Change (IPCC), *Radiative Forcing of Climate Change*, Cambridge University Press, Cambridge, Great Britain (1996).

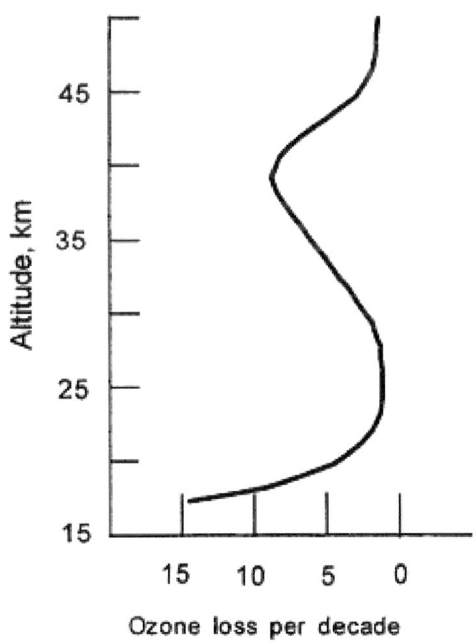

Figure 3.2
Observed trends in vertical distribution of ozone for mid-latitudes (30-50°N) during the 1980s, based on satellite, balloon, and groundbased observations.

concentrations of ozone in regions where there were high concentrations of chlorine oxide, and vice versa, as shown in Figure 3.3. Flights later in September showed even less ozone over Antarctica, as the chlorine continued to react with the stratospheric ozone (see Figure 3.3).

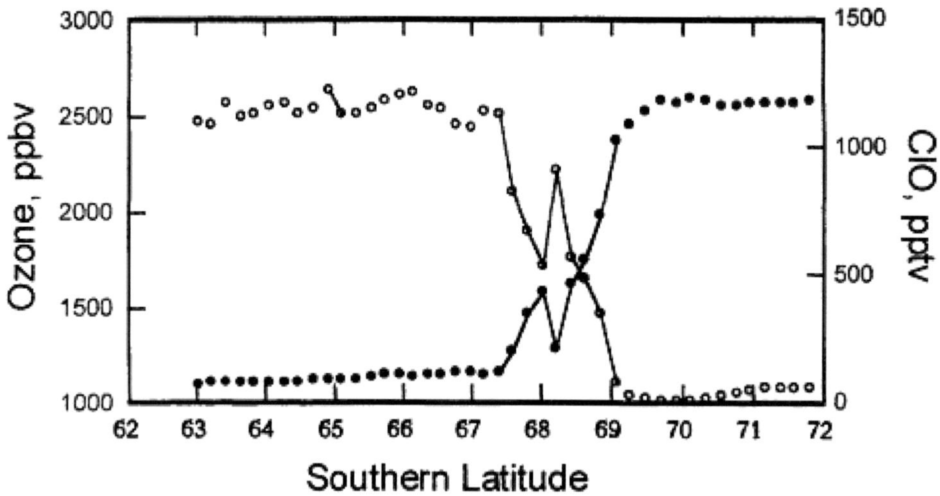

Figure 3.3
NASA aircraft-based measurements of ozone and reactive chlorine (ClO) made in September 1987 over Antarctica. These data demonstrate the relationship between high reactive chlorine and low ozone inside the Antarctic ozone "hole."
Source: National Aeronautics and Space Administration.

Independent measurements made from airplanes, balloons, satellites, and the ground have provided a detailed understanding of the chemical reactions in the Antarctic stratosphere. In the winter, the lower polar stratosphere can reach temperatures sufficiently cold (less than approximately -80°C, or -112°F at 20 m) that stratospheric clouds can form. These clouds facilitate heterogeneous chemical reactions that allow the release of active chlorine in sunlight. The chemical reactions related to the clouds are now well understood through study under laboratory conditions mimicking those found naturally.

The now well-documented phenomenon of the Antarctic ozone hole has demonstrated the cause-and-effect relationship between release of anthropogenic chlorine and bromine and depletion of ozone. Ozone loss has also been observed in the Arctic and, at least in part, is related to processes similar to those occurring in the Antarctic. One of the primary reasons for concern about depletion of stratospheric ozone is the increase in UV radiation that results from the decrease in ozone. Simple physics clearly shows that such UV increases must occur. It is difficult to detect UV changes at the ground caused by small changes in ozone in the stratosphere. In contrast, it is relatively easy to detect the increased UV radiation reaching the surface as a result of the large loss of stratospheric ozone over the Antarctic in springtime. The observations shown in Figure 3.4 agree with the predictions and clearly make the connection between ozone-depleting substances and changes in UV radiation at Earth's surface.

ATMOSPHERIC LIFETIME, OZONE DEPLETION POTENTIAL, AND GLOBAL WARMING POTENTIAL AS INDICATORS OF ENVIRONMENTAL IMPACT

Compounds being considered as replacements for halon include perfluorocarbons, hydrofluorocarbons, hydrochlorofluorocarbons, and several other compounds, such as CF_3I and SF_6. These compounds are listed in Table 3.1, along with their atmospheric lifetimes and calculated ozone depletion potentials. Only individual compounds and their effects are examined here, although the effects of a mixture can be evaluated in terms of the proper ratios of the individual effects. The atmospheric lifetime of such compounds is important in determining their potential effects on ozone and climate, as is their calculated ozone depletion potential (ODP) and global warming potential (GWP).[10]

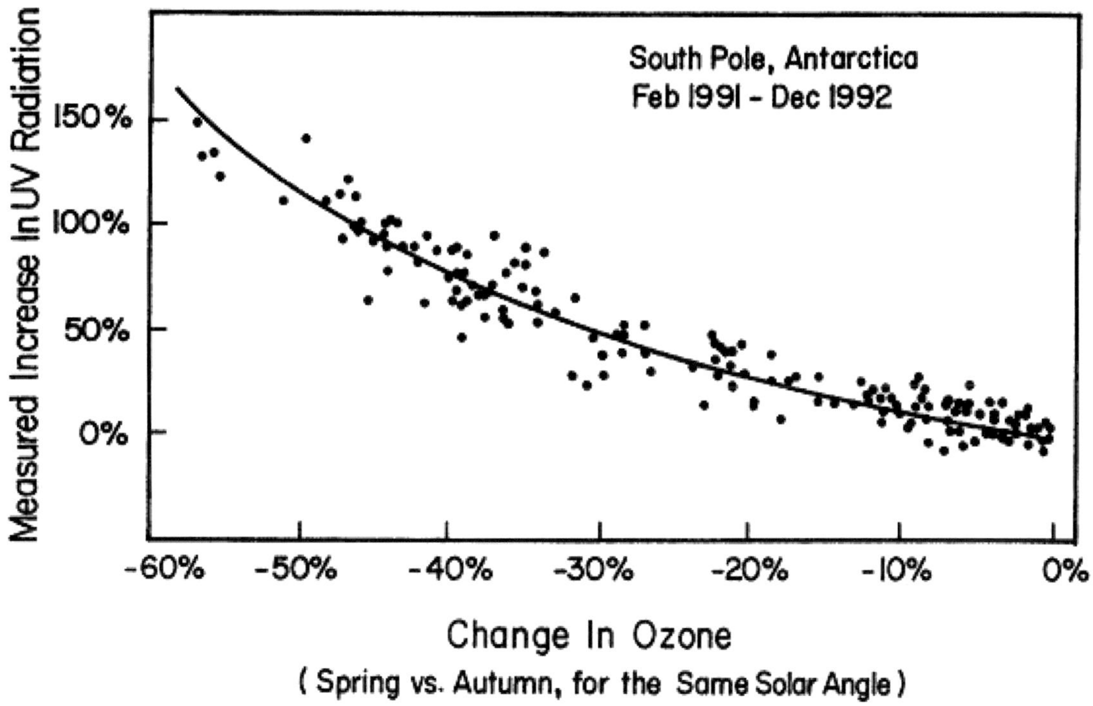

Figure 3.4
Measured increase in erythermal ultraviolet (UV) radiation at the South Pole. These data demonstrate the effects of reduced ozone in the Antarctic ozone "hole" in allowing more UV radiation to reach the surface.
Source: C.R. Booth and S. Mandronich, "Radiation Amplification Factors: Improved Formulation Accounts for Large Increases in Ultraviolet Radiation Associated with Antarctic Ozone Depletion," pp. 39-42 in *Ultraviolet Radiation in Antarctica: Measurements and Biological Research*, C.S. Weiler and P.A. Penhale, Eds., AGU Antarctic Research Series, Vol. 62, American Geophysical Union, Washington, D.C. (1994).

Atmospheric Lifetime

After emission into the atmosphere, the time scale for removal of a gas—its atmospheric lifetime—is generally defined as the ratio of total atmospheric burden to integrated global loss rate. The lifetime is the time it takes for the global amount of the gas to decay to l/e, or 36.8% of its original concentration after initial emission into the atmosphere. The lifetime must take into account all of the processes determining the removal of a gas from the atmosphere, including photochemical losses within the atmosphere (typically due to photodissociation or reaction with OH), heterogeneous removal processes (e.g., loss into clouds or into raindrops), and permanent removal uptake by the land or ocean. Atmospheric lifetimes of a number of gases have been determined based on current knowledge of these loss processes.

As shown in Table 3.1, atmospheric lifetimes of greenhouse gases of interest as replacements for halon range from a few days (e.g., for CF_3I, owing to its photolysis at near-ultraviolet wavelengths) to thousands of years (e.g., for SF_6 and the perfluorocarbons). Lifetimes of HCFCs and HFCs range from very short periods for gases such as HCFC-123 that react rapidly with hydroxyl (OH) to lifetimes comparable in length to those of the halons.

Table 3.1 Atmospheric Lifetime and Calculated Ozone Depletion Potential (ODP) for Halons and for Potential Replacements

Species	Chemical Formula	Atmospheric Lifetime (years)	Ozone Depletion Potential
HALONS			
H-1211	CF_2ClBr	20	5
H-1301	CF_3Br	65	13
H-2402	CF_2BrCF_2Br	20	7
HCFCs			
HCFC-22	CF_2HCl	12.1	0.05
HCFC-123	CF_3CHCl_2	1.4	0.02
HCFC-124	CF_3CHFCl	6.1	0.03
HFCs			
HFC-23	CHF_3	264	~0
HFC-32	CH_2F_2	5.6	0
HFC-125	CF_3CF_2H	32.6	~0
HFC-134a	CH_2FCF_3	14.6	~0
HFC-227ea	CF_3CHFCF_3	36.5	0
HFC-236fa	$CF_3CH_2CF_3$	209	0
PFCs			
FC-14	CF_4	50000	0
FC-116	C_2F_6	10000	0
FC-218	C_3F_8	>2600	0
FC-31-10	C_4F_{10}	>2600	0
FC-51-14	C_6F_{14}	3200	0
OTHER			
CFC-11	$CFCl_3$	50	1
Carbon dioxode	CO_2	*	0
Sulfur hexafluoride	SF_6	3200	0
Trifluoroiodomethane	CF_3I	< 0.005	< 0.008

NOTE: Most of the values for atmospheric lifetime are based on recent international assessments and references therein (IPCC, 1996).
* Decay of CO_2 is a complex function of the carbon cycle (IPCC, 1996).
SOURCE: Adapted from Intergovernmental Panel on Climate Change (IPCC), *Radiative Forcing of Climate Change*, Cambridge University Press, Cambridge, Great Britain (1996).

Potential Effects on Ozone

Understanding of the depletion of stratospheric ozone has led to the need for simple measures for comparing the impact on ozone of different compounds as a scientific guide to public policy. The concept of ozone depletion potential has proven to be a useful index for gauging the effects on ozone from CFCs, halons, and their replacements.

Effectiveness of Chlorine and Bromine in Ozone Destruction

The chlorine and bromine catalytic mechanisms are particularly efficient at destroying ozone. The chlorine and bromine catalytic cycles can occur thousands of times before the catalyst is converted to a less reactive form such as HCl or HBr. Because of this cycling, relatively small concentrations of reactive chlorine or bromine can have a significant impact on the amount and distribution of ozone in the stratosphere. In the lower stratosphere, atmospheric and laboratory measurements indicate that heterogeneous chemistry on particles leads to enhanced effects on ozone from chlorine and bromine by helping to convert less reactive species to the forms of bromine and chlorine that can react catalytically.

Bromine is much more effective at destroying ozone in the lower stratosphere than is chlorine, as much as 100 times more efficient below 20 km.[11,12,13] The emissions and corresponding amount of brominated compounds or halons in the atmosphere are much smaller, however, than those of the chlorinated compounds. As a result, while not negligible, bromine's impact on the current atmosphere is smaller than the effects from increasing chlorine.

CF_3I and other compounds containing iodine have been suggested as replacements for halons. Iodine reaching the stratosphere is even more effective than bromine and over 1,000 times more effective than chlorine at destroying ozone in the lower stratosphere.[14,15]

Other suggested replacements are primarily composed of carbon, fluorine, and hydrogen. None of these compounds affects ozone. It has been suggested that CF_3 radicals produced by dissociation of some of these compounds, such as from the dissociation of HFC-134a, could affect ozone; recent studies show, however, that these radicals do not have any significant effects on ozone.[16,17]

How Much Have Halons Affected Stratospheric Ozone?

The concept of chlorine/bromine loading[18] provides a useful means of examining the relative effects of different chlorine and bromine compounds on stratospheric ozone. By combining atmospheric observations with analyses from atmospheric models, chlorine/bromine loading provides a measure of the amounts of chlorine and bromine available to react with ozone. The changes in chlorine/bromine loading with time, based on past and projected emissions of halocarbons, are assumed to be proportional to changes in ozone using the effects of chlorine and bromine in determining trends in ozone depletion over recent decades. Of particular interest is the chlorine/bromine loading in the lower stratosphere, where observations indicate that most of the ozone loss has occurred.

Figure 3.5 shows the chlorine/bromine loading in the lower stratosphere for known past emissions of CFCs, halons, and other halocarbons and for projected emissions corresponding to those expected following the Copenhagen Agreement modifications to the Montreal Protocol (note that bromine is assumed to be 60 times as effective as chlorine in affecting ozone in this analysis). Taking into account uncertainties associated with the effects of bromine on ozone relative to the effects of chlorine, these analyses suggest that halons account for about 7 to 12% of the ozone decrease in the lower stratosphere through 1995, while total bromine, including effects from human-related emissions of methyl bromine, accounts for 15 to 25% of the ozone decrease in this region. Bromine has almost no effect on determining the ozone decrease in the upper stratosphere, thus implying smaller effects from halons (3 to 5%) and total bromine (9 to 12%) on globally integrated ozone distribution.

Ozone Depletion Potential

The concept of ozone depletion potential (ODP) provides a cumulative measure of the expected effects on ozone from the emissions of a gas relative to one of the gases of most concern with respect to ozone change, namely CFC-11 ($CFCl_3$).[19,20,21,22] This concept is an integral part of national and international policy considerations to protect ozone, including the Montreal Protocol and its amendments, and the U.S. Clean Air Act. ODPs provide an important index for analyzing a new chemical's potential to affect ozone relative to CFCs, halons, or other replacement compounds being considered.

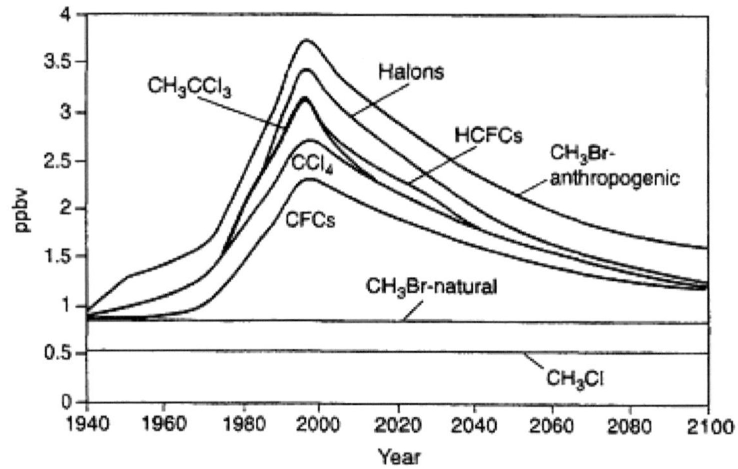

Figure 3.5

Chlorine/bromine loading in the lower stratosphere for known past emissions of CFCs, halons, and other halocarbons and for projected emissions corresponding to those expected following the Copenhagen Agreement modifications to the Montreal Protocol.

The ODP of a gas is defined as the change in total ozone per unit mass emission of the gas relative to the change in total ozone per unit mass emission of CFC-11. Time-dependent ODPs can also be defined that provide information on a compound's effects on ozone on a shorter time scale. However, the steady-state values have been preferred and have been used in regulatory considerations. ODPs are currently determined by two different means: by calculations with two-dimensional models of the global atmosphere[23,24] and by the semiempirical approach developed by Solomon et al.[25] The two approaches give similar results.

By definition, the ODP for CFC-11 ($CFCl_3$) is 1.0, and the calculated ODPs for other banned CFCs are all greater than 0.4. The Clean Air Act calls for policy actions on compounds whose ODPs are equal to or greater than 0.2. As listed in Table 3.1, the steady-state ODPs for halons are all extremely large, much greater than 1.0, reflecting the reactivity of bromine with ozone, whereas the ODPs for all of the hydrofluorocarbons (HFCs), the perfluorocarbons (PFCs), and sulfur hexafluoride are all near zero, reflecting the inability of their degradation products to participate in catalytic ozone destruction cycles.[26] Although the HCFCs do contain chlorine and can affect ozone, the ODPs of those being considered as halon replacements (Table 3.1) are all small, with values of 0.02 to 0.05. The effects on ozone per unit emission of one of these HCFCs would be less than one hundredth of the effect on ozone caused by the halon it would replace. The short lifetimes of these HCFCs and their lack of bromine result in the reduced effect on ozone.

Although iodine is extremely reactive with ozone, the ODP for surface emissions of CF_3I is less than 0.01 (a recent analysis by Connell et al.[27] suggests that the ODP is less than 0.006); because of its reactivity in the troposphere, very little iodine would be expected to reach the stratosphere. However, this ODP value is subject to significant uncertainty because of uncertainty in the understanding of iodine chemistry (e.g., lack of data on such reactions as IO with O_3, ClO, or BrO) and the physical processes (e.g., the effects of fast vertical transport from convective processes in transferring iodine from the lower to the upper troposphere) affecting the iodine distribution in the troposphere and stratosphere.

In evaluating alternatives to halon, time-dependent ODPs are also useful to examine because they provide insight into a compound's short-term effects on stratospheric ozone following its emission (while steady-state ODPs indicate integrated effects over longer time scales). As discussed in recent international assessments,[28] the ODPs for the HCFCs are much larger at short time scales of a few years than they are at steady state. The short atmospheric lifetimes of these compounds imply that, compared to CFC-11, they release chlorine in the lower stratosphere more quickly, and can result in a more immediate (but small) effect on ozone. However, the time-dependent ODPs for these HCFCs are still much smaller than the ODPs for the halons they would replace.

Potential Effects on Climate

Much of the concern about effects of human activities on climate has centered on carbon dioxide (CO_2) because of its importance as a greenhouse gas and also because of the rapid rate at which its atmospheric concentration has been increasing. However, other greenhouse gases account for about half of the overall increase in the effect of radiative forcing on climate.[29] Although halons, because of their small concentrations, are currently only a minor contributor to increased radiative forcing, it is important to consider the possible role of any replacements in affecting future climate.

Several different indices have been used as measures of the strength of the radiative forcing on climate from different greenhouse gases. The indexing approach for greenhouse gases that has gained the widest acceptance is the concept of global warming potential (GWP) originally developed by the Intergovernmental Panel on Climate Change (IPCC).[30]

Radiative Forcing

The radiative forcing of the surface-troposphere system (owing, for example, to a change in greenhouse gas concentration) is defined as the change in net irradiance (in watts per meters squared, Wm^{-2}) at the tropopause after allowing for readjustment of stratospheric temperatures to stratospheric equilibrium. The tropopause is the reference point because it is considered, in a global and annual mean sense, that the surface and the troposphere are closely coupled.

A key factor affecting the radiative forcing associated with a gas is the location of the wavelengths at which it absorbs infrared radiation. The spectral region from about 8 to 12 μm is referred to as the "window" because of the relative transparency of the atmosphere to radiation over these wavelengths. Most of the non-CO_2 greenhouse gases with the potential to affect climate, including halons and most of their replacements, all have strong absorption bands in the atmospheric window region. Relatively small changes in the concentrations of these gases can produce a significant increase in radiative forcing.

As the concentration of a greenhouse gas becomes high, it can absorb most of the radiation in its energy bands; once any of its absorption wavelengths become saturated, it is unable to absorb more energy at those wavelengths, and a further increase in its concentration has a diminishing effect on climate. This is called the band saturation effect. For example, the radiative forcing attributable to further increases in carbon dioxide concentrations in the current atmosphere will increase as the natural logarithm of its concentration because of this effect. Also, at the wavelengths where water vapor and carbon dioxide strongly absorb infrared radiation, the greenhouse effect of other gases will be minimal. However, absorption by other gases, such as the halons or other halocarbons, at wavelengths that are not saturated varies linearly with concentration. Another important consideration in radiative absorption is the band overlap effect. If a gas absorbs at wavelengths that are also absorbed by other gases, then the effect on radiative forcing of increasing its concentration can be diminished.

In addition to the direct forcing effect from emission of a gas into the atmosphere, the net radiative forcing can also be modified through indirect effects relating to chemical interactions on other radiatively important constituents. For example, emissions of halons result in stratospheric bromine that can destroy stratospheric ozone, which is also a greenhouse gas.[31] Such indirect effects need to be considered when candidate replacement compounds are evaluated for potential effects on climate.

Table 3.2 Radiative Forcing Due to Halon 1301 and Potential Replacements

Species	Chemical Formula	Radiative Forcing per Unit Mass Relative to CFC-11 ($CFCl_3$)	Radiative Forcing per Unit Molecule Relative to CFC-11 ($CFCl_3$)
HALON			
H-1301	CF_3Br	1.19	1.29
HCFCs			
HCFC-22	CF_2HCl	1.37	0.86
HCFC-123	CF_3CHCl_2	0.72	0.80
HCFC-124	CF_3CHFCl	0.88	0.87
HFCs			
HFC-23	CHF_3	1.59	0.81
HFC-32	CH_2F_2	1.06	0.40
HFC-125	CF_3CF_2H	1.03	0.90
HFC-134a	CH_2FCF_3	1.04	0.77
HFC-227ea	CF_3CHFCF_3	0.95	1.17
HFC-236fa	$CF_3CH_2CF_3$	1.06	1.17
PFCs			
FC-14	CF_4	0.69	0.44
FC-116	C_2F_6	1.36	1.37
FC-218	C_3F_8	0.77	1.05
FC-51-14	C_6F_{14}	0.75	1.84
OTHER			
CFC-11	$CFCl_3$	1	1
CO_2	CO_2	8×10^{-5}	2.5×10^{-4}
Sulfur hexafluoride	SF_6	2.75	2.92
Trifluoroiodomethane	CF_3I	1.20	1.71

NOTE: The absolute radiative forcing due to CFC-11 is 0.22 Wm^{-2} per ppbv increase (Intergovernmental Panel on Climate Change (IPCC), *Radiative Forcing of Climate Change*, Cambridge University Press, Cambridge, Great Britain, 1996). The radiative forcing for CFC-11 is a factor of 3,970 greater than that for CO_2 per unit mass and 12,400 greater than that for CO_2 per unit molecule increase.
SOURCE: Adapted from IPCC, 1996, and from World Meteorological Organization, *Scientific Assessment of Ozone Depletion*, 1994.

Relative Radiative Forcing per Molecule or Mass

Relative radiative forcing represents a comparison of radiative forcing on a molecule-per-molecule or kilogram-per-kilogram basis for the different greenhouse gases. It is generally given relative to CO_2 or CFC-11. A radiative transfer model of the atmosphere is used to determine the radiative forcing attributable to small perturbations of these gases relative to present-day conditions. Small perturbations are used in the calculations in order to prevent the marked nonlinear absorption of carbon dioxide, methane, and nitrous oxide from affecting the radiative forcing for these gases. Table 3.2 shows radiative forcing on a per-molecule basis relative to CFC-11 for halons and their possible replacements.

The values for radiative forcing for halon 1301 and possible replacements listed in Table 3.2 are all within a factor of four of each other and are roughly of the same order of magnitude as the values attributable to CFC-11. On the other hand, the radiative forcing for CFC-11 is 3,970 times greater than that for CO_2 on a per-unit-mass basis and is 12,400 times greater than that for CO_2 on a per-molecule basis. These results suggest that all of the replacement compounds could be important greenhouse gases if their atmospheric concentrations became large enough.

What Effects Could Replacements for Halons Have on Climatic Change?

Current concentrations of halon 1301 in the atmosphere (about 0.003 ppbv) produce a radiative forcing on climate that is 0.0005 that owing to increases in carbon dioxide concentrations over the last two centuries. Even over recent decades the effect on climate of the rapid increases in halon concentrations would still be small relative to the effect of the increase in CO_2 concentration. Increased halon concentrations due to use by the U.S. Navy account for an even smaller fraction of the radiative forcing effects on climate.

Global Warming Potential

Akin to the concept of ODP, the concept of global warming potential has been developed to provide a simple representation of the relative effects on climate resulting from a unit-mass emission of a greenhouse gas. Values for GWP are expressed as the time-integrated radiative forcing from the instantaneous release of a kilogram (i.e., a small mass emission) of a trace gas expressed relative to that of a kilogram of the reference gas, CO_2:

$$GWP(x) = \frac{\int_0^n F([x(t)])dt}{\int_0^n F([CO_2(t)])dt} , \qquad (1)$$

where n is the time horizon over which the calculation is considered; $F([x(t)]$ is the radiative forcing in response to the changing concentration of species x after the pulse emission at time $t = 0$; $[x(t)]$ is the time-decaying concentration of that gas; and the corresponding quantities for the reference gas are in the denominator. The radiative forcing responses are derived from radiative transfer models. The trace gas concentrations, $[x(t)]$, remaining after time t are based on the atmospheric lifetimes of the gas in question. The reference gas has been taken generally to be CO_2, since this allows a comparison of the radiative forcing role of the emission of the gas in question to that of the dominant greenhouse gas that is emitted as a result of human activities.

The best choice of integration time horizon in evaluating GWP has been the subject of much discussion and controversy. Unlike the case with ODP, the complexities of treating CO_2 and the carbon cycle prevent integration of GWP to steady state. There is, however, no given value of integration time for determining GWP that is ideal over the range of uses of this concept. Values for GWP are generally calculated over three time horizons, for 20, 100, and 500 years. It is believed that these three time horizons provide a practical range for policy applications.[32] GWPs determined for the longest integration period provide a measure of the cumulative chronic effects on climate, while integration over the shortest period is representative of near-term effects. GWPs evaluated over the intermediate 100-year period appear generally to provide a balanced representation of the various time scales for climatic response. The best choice of time horizon will depend on the specific analysis being considered. It is necessary to balance the effects of near-term responses in comparing greenhouse gases with consideration of the long-term persistence of any environmental effects from long-lived gases.

Table 3.3 Direct Global Warming Potential of Halon 1301 and of Candidate Replacements

	Chemical Formula	Atmospheric Lifetime (years)	Global Warming Potential (Time Horizon)		
Species			20 years	100 years	500 years
CO_2	CO_2	*	1	1	1
HALON					
H-1301	CF_3Br	65	6100	5400	2200
HCFCs					
HCFC-22	CF_2HCl	12.1	4000	1500	520
HCFC-123	CF_3CHCl_2	1.4	300	90	29
HCFC-124	CF_3CHFCl	6.1	1500	470	140
HFCs					
HFC-23	CHF_3	264	9100	11700	9800
HFC-32	CH_2F_2	5.6	2100	650	200
HFC-125	CF_3CF_2H	32.6	4600	2800	920
HFC-134a	CH_2FCF_3	14.6	3400	1300	420
HFC-227ea	CF_3CHFCF_3	36.5	4300	2900	950
HFC-236fa	$CF_3CH_2CF_3$	209	5100	6300	4700
PFCs					
FC-14	CF_4	50000	4400	6500	10000
FC-116	C_2F_6	10000	6200	9200	14000
FC-51-14	C_6F_{14}	3200	5000	7400	10700
OTHER					
CFC-11	$CFCl_3$	50	4900	3800	1300
Sulfur hexafluoride	SF_6	3200	16300	23900	34900
Trifluoroiodomethane	CF_3I	<0.005	<3	<<1	<<<1

[*] Decay of CO_2 is a complex function of the carbon cycle (IPCC, 1996).
SOURCE: Adapted from Intergovernmental Panel on Climate Change (IPCC), *Radiative Forcing of Climate Change*, Cambridge University Press, Cambridge, Great Britain (1996).

Direct GWPs Table 3.3 summarizes the GWPs for halons and their replacements included in the new IPCC assessments. Because their radiative forcing is of similar magnitude, the replacements with the longer atmospheric lifetimes have the largest GWPs. In fact, the GWPs for many of the replacements are as large as or larger than the GWPs for halon 1301 (other halons have not been evaluated), particularly for the 20-year integration time. The shorter-lived compounds, such as HCFC-123, HCFC-124, and HFC-32, have appreciably smaller GWPs, particularly at the 100- and 500-year integration periods. The GWPs for CF_3I are extremely small (<3 even for the 20-year integration) owing to its atmospheric lifetime of only a few days.

The GWPs for the perfluorocarbons and for SF_6 are all much larger than any of the GWP values for CFCs or halons. The very long atmospheric lifetimes of these gases lead to extremely large GWPs. These large GWPs imply potentially large effects on climate over long time scales, with the actual effect on climatic radiative forcing dependent on the magnitude of emissions into the atmosphere.

Indirect GWPs Possible effects on ozone from emissions of halons or their replacements also need to be considered. Daniel et al.[33] have estimated the indirect GWPs for effects on ozone from a variety of halocarbons, including CFCs, halons, and HCFCs, in an attempt to clarify the relative radiative forcing roles of different classes of ozone-depleting compounds. Decreased ozone from CFCs and halons should decrease the radiative forcing on climate. Daniel et al. found that the net GWP of halocarbons depends strongly on the each compound's effectiveness for ozone destruction. Halons are likely to have negative net GWPs, while those of CFCs are likely to be positive over both 20- and 100-year time horizons. These analyses, however, are still subject to remaining uncertainties about the causes of ozone decreases in the lower stratosphere. These indirect effects are not included in the values given in Table 3.3.

Summary

The concepts of ozone depletion potential and global warming potential provide important guides to the potential of halon replacements to destroy stratospheric ozone and to affect radiative forcing on climate. In general, unless the amounts of replacement compounds produced and emitted into the atmosphere are much larger than the amounts of the compounds they replace, the compounds being considered as halon replacements are not expected to have any appreciable effect on stratospheric ozone.

However, the perfluorocarbons and SF_6 have extremely long atmospheric lifetimes (> 1,000 years) and their large GWPs suggest that emissions of these compounds could lead to significant concerns about radiative forcing on climate.

OTHER ENVIRONMENTAL EFFECTS

A chemical compound introduced into the atmosphere at the ground will be dispersed throughout the troposphere. It takes a few months for the compound to be dispersed within the hemisphere in which it is released. The time constant for cross-equatorial transport is about 1 year. On the same time scale, a portion of the chemical will also be transported to the stratosphere. Material is continuously recycled between the troposphere and the stratosphere. An inert molecule spends 90% of its time in the troposphere. Once transported to the stratosphere, a compound has an average residence time of about 3 years before it is transported back to the troposphere.

The extent to which a compound can be dispersed globally depends on the competition between transport and removal processes. Removal from the atmosphere is achieved by chemical or photochemical reactions that alter the identity of the compound, or by physical means such as rainout. Figure 3.6 provides a schematic representation of the life cycle of a chemical, *A*, released to the atmosphere. It is dispersed between the ground, tropospheric, and stratospheric reservoirs. Possibilities exist for the compound to be transformed into an intermediate product, *B*, and/or final products *C* and *D* in each reservoir.

If the compound is insoluble and chemically inert, the material will retain its identity as *A*, and all of it will stay in the atmosphere. Eventually, the material will be dispersed evenly throughout the atmosphere with a constant volume mixing ratio.* One concern in this case is that the concentration of *A* in the atmosphere would reach a sufficient level that it could block infrared radiation emitted from the surface, leading to greenhouse warming. Given the radiative properties of most molecules, the greenhouse warming effect starts to become significant when a few teragrams (1 teragram = 10^{12} gm) of the compound are present in the atmosphere so that its mixing ratio approaches the parts per billion by volume level. The GWP of a compound provides a relative measure of its greenhouse warming impact on a unit mass emitted basis (see discussion in preceding section).

A more typical situation is presented by a chemical whose reactivity in the atmosphere is sufficiently slow (removal rate slower than 1% per day) that it can be transported globally before it is chemically

* The volume mixing ratio of a species *A* is the ratio of the number of *A* molecules in an air sample to the total number of air molecules in the same sample. For a trace gas with constant mixing ratio, any sample of air will contain the same fraction of *A*.

removed. The final degradation products (*C*) for the halocarbons are CO_2 from the carbon atoms, and hydrogen halides (i.e., HF, HCl, HBr, and HI) from the halogen atoms. The amount of CO_2 produced by the oxidation of halocarbons is negligible when compared to the magnitude of the other CO_2 sources (e.g., fossil fuel combustion). The hydrogen halides are removed from the atmosphere by rainout. The resulting acid deposition is negligible compared to that from other sources present in the atmosphere.[34] With the exception of fluorine, the halogen atoms (i.e., Cl, Br, and I) coexist with the hydrogen halides in the atmosphere. Cl, Br, and I atoms participate in catalytic chemical cycles that can deplete stratospheric ozone; F atoms do not. F atoms react rapidly with H_2O vapor and with CH_4 to give HF. The H-F bond is extremely strong. HF is an efficient and permanent sink for F atoms in the atmosphere, and as a result F atoms do not participate in catalytic chemical cycles.[35] The value of the ozone depletion potential assigned to a compound is a combined measure of the amount of free radicals that it delivers to the stratosphere and the catalytic efficiency of the free radicals in ozone removal. ODPs measure the relative effects of different compounds on ozone on a per unit mass emitted basis, not on the basis of absolute concentrations oft he halogen radicals delivered to the stratosphere. For a detailed discussion of ODPs and their derivation, see the discussion in the preceding section.

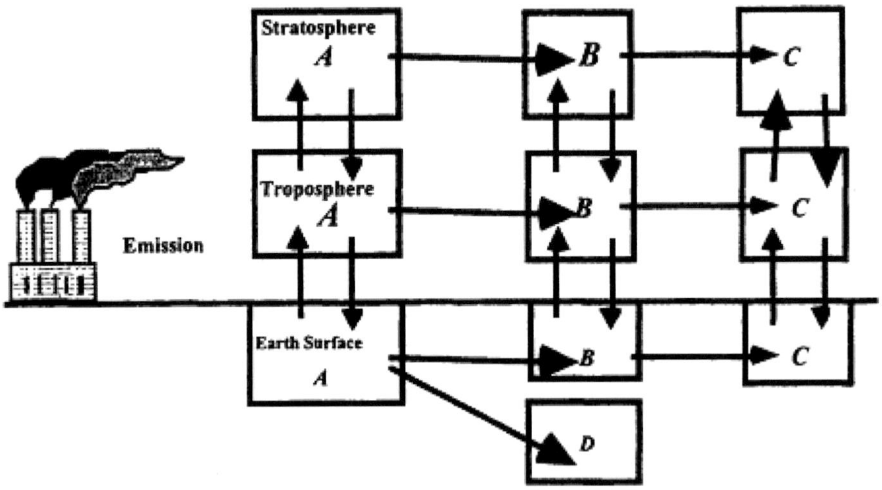

Figure 3.6
Schematic representation of a chemical's life cycle. The horizontal arrows represent chemical and/or physical processes. Vertical arrows represent transport between reservoirs.

Intermediate products may also be formed during the atmospheric degradation of halon replacements. The question addressed here is whether any of these products will have any significant environmental impact. There are three issues to be considered:

1. If the atmospheric concentrations of the intermediate products (*B*) and final products (*C*) are sufficiently high, their greenhouse warming effects should be added to the global warming potential of the parent compound.
2. The intermediate products may react with other trace gases in the atmosphere and perturb their natural balance in the atmosphere. The effect on ozone is of particular interest.
3. The ecological effects of the deposition of intermediate and final products on the ground should be assessed. A necessary step in assessing these effects is the estimation of the concentrations of each chemical and their by-products in each reservoir.

Estimating Atmospheric Concentrations of Intermediate Degradation Products

The atmospheric concentration of a substance depends on the rate at which it is emitted to, or produced in situ, in the atmosphere and the rate at which it is removed from the atmosphere. The local chemical removal rate can be characterized by a time constant that is related to the inverse of the reaction rate. For two species with the same emission or production rates, the one with the longer lifetime will have a larger concentration in the atmosphere. For all of the CFCs, HCFCs, and HFCs that have been examined, the lifetimes of the intermediate atmospheric degradation products are always much shorter than those of the parent compounds. As a result, the atmospheric concentrations of the intermediate products are always much smaller than those of the parent compounds.

Chlorofluorocarbons (CFCs) and halons are essentially inert in the troposphere but are reactive in the stratosphere. Intermediate degradation products from CFCs and halons usually have small concentrations in the stratosphere because there are abundant high-energy UV photons in the stratosphere to dissociate them. The initial sequence of reactions leads rapidly (within a few minutes) to the formation of intermediate oxidation products. One example of an intermediate product is CF_2O from CFC-12 (CF_2Cl_2). CF_2O is further degraded in the atmosphere to form HF and CO_2. The atmospheric concentration of CF_2O is expected to be about a factor of 1,000 less than that of CFC-12. Concentrations of CF_2O observed in the atmosphere are in agreement with expected concentrations calculated using known photochemical processes.[36]

Hydrochlorofluorocarbons (HCFCs), hydrofluorocarbons (HFCs), and hydrogenated halons react with the hydroxyl radical in the troposphere. Although their primary loss is in the stratosphere, some halons (1211, 1202, and 2402) do have a small fraction of their photodissociation in the troposphere. Intermediate products produced in the troposphere from the degradation of HCFCs and HFCs are subject to removal by rainout within a few day[37] As a result, the concentrations of the intermediate products in the atmosphere are very small.

Estimating the Concentrations of the Intermediate Products on the Ground

Some of the intermediate degradation products produced in the troposphere may be removed and deposited back to the ground by rainout and dry deposition before they can be photochemically converted into CO_2 and the respective halogen halides. A simple estimate of the upper-limit rainwater concentration of the intermediate product can be obtained as follows. If one assumes that 1 kiloton of compound in the atmosphere would produce 1 kiloton of the intermediate degradation product, the annual total rainfall of 5×10^{17} liter would imply an average concentration of the intermediate product at 2 ng/liter for a 1 kiloton/yr emission. More sophisticated estimates can be obtained by taking into account the yield of the intermediate product, the spatial distribution of the gas-phase precursors, and the distribution of precipitation. This can be achieved using a three-dimensional chemistry-transport model. An example of such a study of the global-scale deposition of CF_3COOH (TFA) produced during the atmospheric oxidation of HFC-134a can be found in Kanakidou et al.[38] In addition, regional-scale acid deposition models can be used to examine the variations on local and regional scales.

Once the rainwater reaches the ground, one needs to assess if the intermediate product can be degraded in the natural environment, e.g., by microbial action in the ground water. Tromp et al.[39] pointed out that in certain water bodies, the normal evaporation cycle will remove the water, leaving behind the solute. In the absence of a degradation mechanism, the solute could accumulate to concentrations that are much higher than those found in rainwater.

Degradation Pathways for Candidate Alternative Fire Extinguishing Agents

The halon alternatives that have been identified consist of a number of HCFCs, HFCs, and PFCs and CF_3I. They can be separated into the following groups:

• Halogenated methanes:	HCFC-22 ($CHClF_2$), HFC-23 (CF_3H), IFC-1211 (CF_3I)
• Halogenated ethanes:	HCFC-123 (CF_3CHCl_2), HCFC-124 (CF_3CHClF), HFC-125 (CF_3CHF_2), HFC-134a (CF_3CH_2F)

• Halogenated propanes:	HFC-227ea (CF_3CHFCF_3), HFC-236fa ($CF_3CH_2CF_3$)
• Perfluorinated halocarbons:	PFC-218 ($CF_3CF_2CF_3$), PFC-31-10 ($CF_3CF_2CF_2CF_3$).

Degradation Pathways for Halomethanes and Haloethanes

The atmospheric degradation mechanisms of halomethanes and haloethanes have been the subject of several workshops. The results are available in a 1990 WMO report.[40]

Atmospheric degradation of halomethanes and haloethanes is initiated principally by the reaction with the OH radicals. The products are given in Table 3.4. The degradation pathways for the typical haloethane are illustrated in Figure 3.7, after Cox and Lesclaux[41] and Wallington et al.[42] The detailed discussion that follows focuses on haloethane of the form CX_3CYZH, where X = Cl or F, and Y and Z can be H, Cl, or F. The HFC/HCFC molecule reacts with OH or $O(^1D)$, giving rise to haloalkyl (R) radicals, labeled A in Figure 3.7. Haloalkyl radicals add molecular oxygen rapidly (within 1 microsecond) to give haloalkyl peroxy radicals, labeled B_2. Haloalkyl peroxy radicals undergo a variety of reactions and within a few minutes are converted into hydroperoxides (ROOH), labeled B_1; alkyl peroxynitrates ($ROONO_2$), labeled B_3; and alkoxy (RO) radicals, labeled C. As shown in Figure 3.7, a variety of processes (reaction with OH radicals, thermal decomposition, and photolysis) return the hydroperoxides and peroxynitrates to the peroxy and alkoxy radical pool. While the atmospheric lifetimes of hydroperoxides and peroxynitrates vary depending on their chemical identity and the region of the atmosphere in which they are formed, these compounds typically have lifetimes in the range of several hours to several days. Haloalkoxy radicals are reactive species and have lifetimes that are less than 1 second. There are several possible degradation pathways for the haloalkoxy radicals. Alkoxy radicals can decompose by two different bond cleavage routes, C-C bond cleavage (path P1) to form fragments labeled D_1, or C-Cl cleavage (path P3) to form the acetyl halide (CX_3CYO), labeled D_2. For alkoxy radicals containing an -H atom, reaction with O_2 (path P2) to form an acetyl halide is a possible atmospheric fate. The CX_3CF_2O radicals undergo C-C bond cleavage via path P1 exclusively. CF_3CCl_2O and CF_3CFClO radicals decompose by C-Cl bond cleavage (P3) exclusively. For CF_3CHFO radicals there are two important pathways. The first is C-C bond scission and the second is reaction with

Table 3.4 Atmospheric Degradation Products of Selected Fluorocarbons

Fluorocarbon	Structural Formula	Intermediate Degradation Products	Final Degradation Products
HCFC-22	$CHClF_2$	CF_2O	HF, HCl, CO_2
HFC-23	CF_3H	CF_2O, CF_3OH	HF, CO_2
IFC-13I1	CF_3I	CF_2O, CF_3OH	HF, HI, CO_2
HCFC-123	CF_3CHCl_2	$CF_3C(O)Cl$	HCl, CF_3COOH
HCFC-124	CF_3CHClF	$CF_3C(O)F$	HCl, CF_3COOH
HFC-125	CF_3CHF_2	CF_2O, CF_3OH	HF, CO_2
HFC-134a	CF_3CH_2F	HC(O)F, CF_3OH, $CF_3C(O)F$, CF_2O	HF, CF_3COOH, CO_2
HFC-227ea	$CF_3CHF\ CF_3$	CF_3OH, $CF_3C(O)F$, CF_2O	HE, CF_3COOH, CO_2
HFC-236fa	$CF_3CH_2\ CF_3$	$CF_3CO\ CF_3$, CF_3OH, CF_2O, CO	HF, CO_2
PFC-218	$CF_3CF_2\ CF_3$		HF, CO_2
PFC-31-10	$CF_3CF_2CF_2\ CF_3$		HF, CO_2

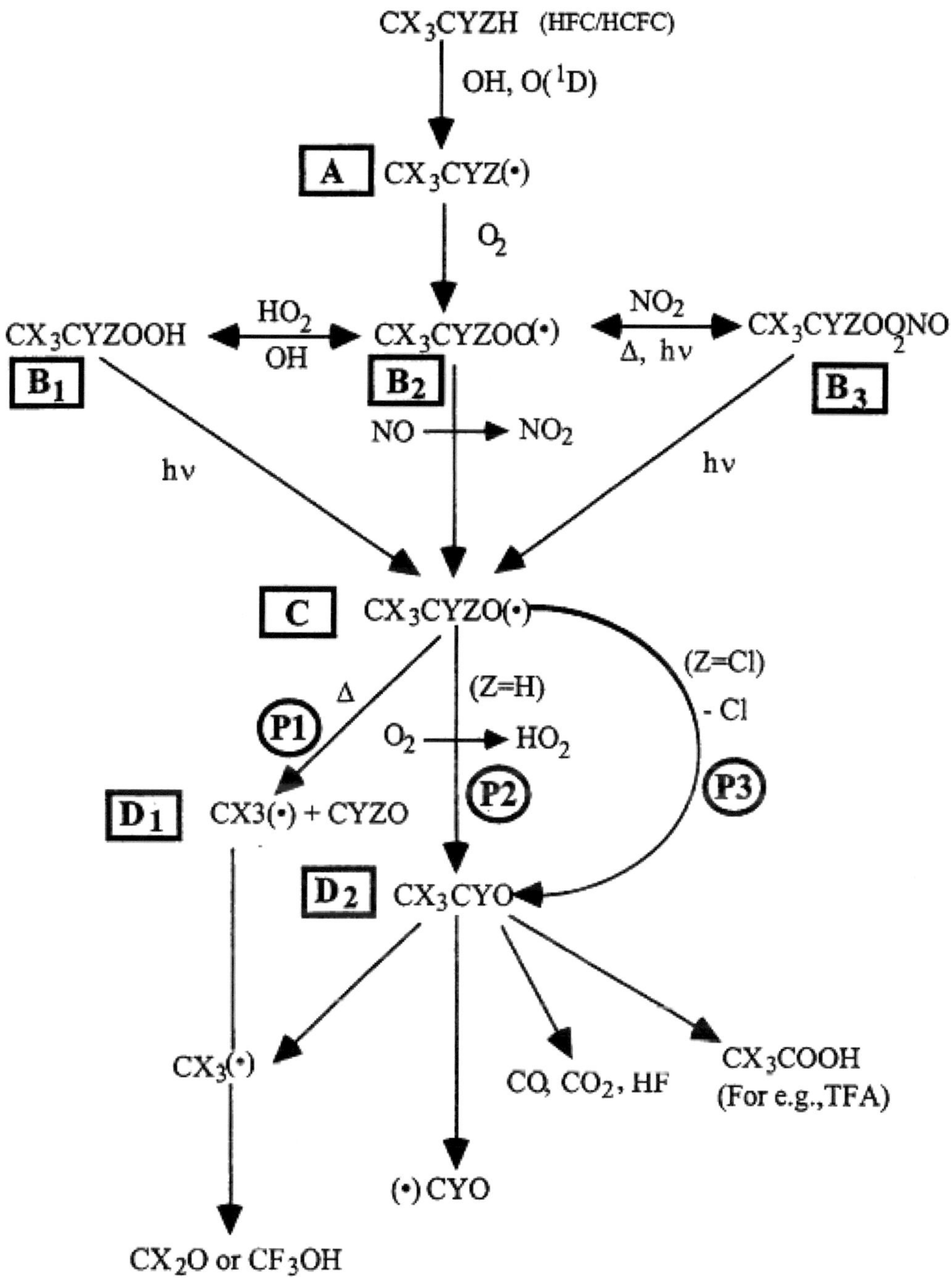

Figure 3.7

Degradation pathways for haloethane of the form CX_3CYZH, where X = Cl or F, and Y and Z can take the form of H, Cl, or F. The symbol "(•)" denotes a radical species, and D denotes thermal decomposition. Source: Adapted from R.A. Cox and R. Lesclaux, "Degradation Mechanisms of Selected Hydrochlorofluorocarbons in the Atmosphere: An Assessment of Current Knowledge" in *Scientific Assessment of Stratospheric Ozone: 1989*, Vol. II, World Meteorological Organization, Global Ozone Research and Monitoring Project—Report No. 20, Geneva (1990); and T.J. Wallington, W.F. Schneider, D.R. Worsnop, O.J. Nielsen, J. Sehested, W.J. Debruyn, and J.A. Shorter, "Environmental Impact of CFC Replacements—HFCs and HCFCs," *Environ. Sci. Technol.* 28, 320A-326A (1994).

O_2. The relative importance of these two processes is dependent on temperature, O_2 partial pressure, and total pressure. Because of this dependence, the branching will vary with altitude. For example, recent laboratory measurements[43,44] for the CF_3CHFO radical decomposition indicate that approximately 30% of these radicals will degrade to $CF_3C(O)F$. CX_3CH_2O radicals react with O_2 to form the aldehyde, CX_3 CHO, and HO_2 radicals. The aldehydes and acid halides are soluble in water and will then react further to form halogenated acids or hydrogen halides and carbon dioxide. The probable degradation products for the haloethanes are summarized in Table 3.4.

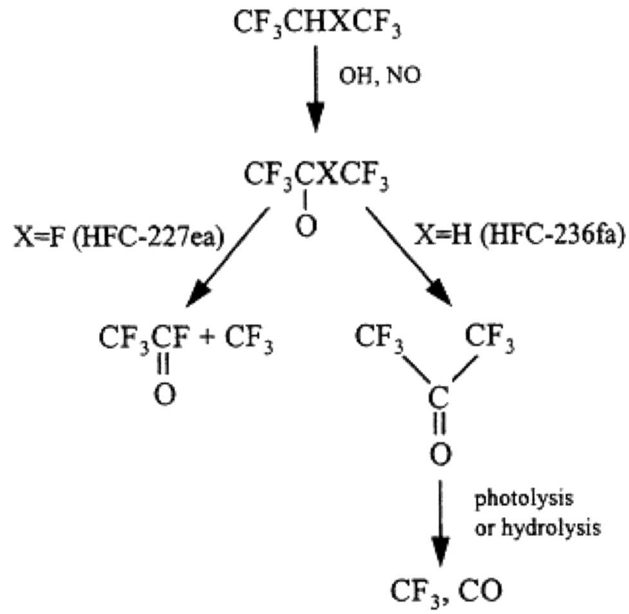

Figure 3.8
Degradation scheme for halopropanes.

In the atmosphere CF_3 radicals form CF_3O radicals, which react with hydrocarbons and NO to give CF_3OH and COF_2, respectively.[45] The atmospheric fate of COF_2, CF_3OH, and the acetyl halides is dominated by incorporation into cloud-rain-seawater followed by hydrolysis. Hydrolysis of CF_3COF (produced by HCFC-123, HCFC-124, and HFC-134a) gives trifluoroacetic acid, CF_3COOH. Hydrolysis of COF_2 and CF_3OH gives HF and CO_2, which, at the concentrations expected from HFC degradation, are of no environmental concern.

Degradation Schemes for Halopropanes

Two partially fluorinated propanes are under consideration as halon replacements: HFC-227ea (CF_3CHFCF_3) and HFC-236fa ($CF_3CH_2CF_3$).

Atmospheric oxidation of HFC-227ea gives rise to $CF_3C(O)F$ molecules and CF_3 radicals (Figure 3.8).[46,47] The fate of the CF_3 fragments is discussed in the section immediately above. The atmospheric fate of $CF_3C(O)F$ is incorporation into rain-cloud-seawater, followed by hydrolysis to give trifluoroacetic acid, CF_3COOH.

Atmospheric oxidation of HFC-236fa gives hexafluoroacetone (CF_3COCF_3).[48] The fate of CF_3COCF_3 is either photolysis or interaction with water surfaces. While the relative importance of these two processes has not been measured, it seems likely that they are of comparable importance. The lifetime of CF_3COCF_3 with respect to photolysis has been estimated to be 3 days.[49] Photolysis will generate CF_3 radicals and CO. CF_3 radicals are converted into CF_3O radicals that are converted into

COF_2, which is then hydrolyzed to give HF and CO_2. Hydrolysis of CF_3COCF_3 produces the sesquihydrate $CF_3C(OH)_2CF_3$, the fate of which is unknown.

Atmospheric Degradation of Perfluorocarbons

The atmospheric degradation of perfluorocarbons (PFCs; CF_4, C_2F_6, C_3F_8, C_4F_{10}, etc.) has been investigated by Ravishankara et al.[50] The usual mechanisms by which organic compounds are removed from the atmosphere —such as reaction with OH radicals and with $O(^3P)$ atoms, and photolysis in the lower atmosphere—are of no importance for perfluorocarbons. As a result, perfluorocarbons have extremely long atmospheric lifetimes (2,000 to 50,000 years). Ravishankara et al.[51] reported that the most important removal mechanism for perfluorocarbons is photolysis by absorption of Lyman- vacuum UV radiation at 121.6 nm at high altitudes (> 60 km) in the mesosphere and thermosphere. Two perfluorocarbons are currently under consideration as halon replacements: PFC-218 (C_3F_8), and PFC-31-10 (n-C_4F_{10}). By analogy to the existing database for other perfluorocarbons, it is anticipated that the atmospheric lifetime of these compounds is 2,000 to 3,000 years.[52] As discussed above, the long atmospheric lifetimes of PFC-218 and PFC-31-10 lead to high GWP values for these species. Photolysis of PFC-218 and PFC-31-10 will give CF_3, C_2F_5, and C_3F_7 radicals. These radicals are the same as those produced in the atmospheric degradation of HFC-23 (CF_3H), HFC-125 (C_2F_5H), and HFC-227ca ($C_2F_5CF_2H$), and their atmospheric transformation into the final products shown in Table 3.4 is well established. There are no known closed-shell (i.e., non-radical) intermediate degradation products of PFCs.

Assessment of the Environmental Impacts—ODP and GWP—of the Intermediate Products

Since the ODP and GWP are relative measures on a per unit mass basis, it is important to look at the atmospheric concentration of a degradation product relative to that of its respective parent compound. There appears to be no long-lived intermediate that will lead to high concentration in the atmosphere. While there has been one report that in the presence of O_2 microbial degradation of trifluoroacetic acid (CF_3COOH) produces CHF_3,[53] which has a lifetime of several hundred years, other workers have been unable to reproduce this result.[54] The available evidence shows that the contributions to indirect GWP of the atmospheric degradation products of HFCs, HCFCs, and PFCs are negligible compared to the direct GWP of the parent compounds.

There have been suggestions that CF_3O_x radicals formed by CF_3 fragments may react with ozone in catalytic cycles, leading to depletion of stratospheric ozone. However, recent modeling and laboratory studies[55,56,57] and work by Wallington et al.[58] have shown that there is no impact of CF_3O_x, or any other HFC degradation intermediate, on stratospheric ozone. As with HFCs, perfluoroalkanes have no impact on stratospheric ozone.[59]

Questions have also been raised as to whether intermediate products (if they contain chlorine and bromine atoms) can act as carriers of additional chlorine/bromine atoms from the troposphere to the stratosphere. These questions have been answered by recent experimental studies that have shown that the lifetimes of the intermediate products are much shorter than the transport time to the stratosphere, and hence the intermediate oxidation products do not act as carriers of chlorine/bromine atoms to the stratosphere.

Accumulation of Degradation Products in the Biosphere

Given the expected emission rates and the kinetic data available, it is unlikely that the concentrations of the intermediate products in the global atmosphere will approach the parts per trillion level. Hydrolysis of COF_2 and CF_3OH gives HF and CO_2, which, at the concentrations expected from HFC degradation, are of no environmental concern. Trifluoroacetic acid (CF_3COOH) has been identified as a degradation product in several of the candidate agents for fire extinguishment Thus far, no biotic or abiotic destruction mechanism has been clearly established for CF_3COOH, and consequently it is thought to be quite stable in water.[60] CF_3COOH is not toxic toward animals but does have a mild herbicidal effect.[61] The global average concentration of CF_3COOH expected in rainfall as a result of the atmospheric degradation of HFCs is many orders of magnitude below that observed to have an impact on plant systems.[62] However, uncertainties concerning the persistence of CF_3COOH in ground water allow for the possibility that CF_3COOH could accumulate in seasonal wetlands over long time periods (≈100 years).[63] Very recently, Frank et al.[64] have reported the detection of CF_3COOH in lake, river, spring, and ocean water samples at levels which are orders of magnitude greater than can be accounted for by man-made emissions. The work of Frank et al. implies that there are large unknown sources of CF_3COOH. It seems unlikely that the concentration of CF_3COOH formed during the atmospheric degradation of the halon replacement compounds will have any significant adverse global environmental impact. Research is needed to better establish the environmental inventory and fate of CF_3COOH.

REFERENCES

1. CIAP Monograph 3, The Stratosphere Perturbed by Propulsion Effluents, U.S. Department of Transportation, Washington, D.C. (1975).
2. P.J. Crutzen. "The Influence of Nitrogen Oxides on the Atmospheric Ozone Content," *Q. J. R. Meteorol. Soc.* **96,** 320-325 (1970).
3. F.S. Rowland and M.J. Molina, "Chlorofluoromethanes in the Environment," *Rev. Geophys. Space Phys.* **13,** 1-36 (1975).
4. World Meteorological Organization, *Scientific Assessment of Ozone Depletion* (1994).
5. R. Stolarski, R. Bojkov, L. Bishop, J. Staehelin, and J. Zawodny, "Measured Trends in Stratospheric Ozone," *Science* **256,** 342-349 (1992).
6. World Meteorological Organization, *Scientific Assessment of Ozone Depletion* (1994); World Meteorological Organization, Global Ozone Research and Monitoring Project, Report No. 37D, Geneva (1995).
7. World Meteorological Organization, *Scientific Assessment of Ozone Depletion* (1994); World Meteorological Organization, Global Ozone Research and Monitoring Project, Report No. 37D, Geneva (1995).
8. J.C. Farman, B.G. Gardiner, and J.D. Shanklin, "Large Losses of Total Ozone in Antarctica Reveal Seasonal ClO_x/NO_x Interaction," *Nature* **315,** 207-210 (1985).
9. J.G. Anderson, D.W. Toohey, et al. "Free Radicals Within the Antarctic Vortex: The Role of CFCs in Antarctic Ozone Loss," *Science* **251,** 39-46 (1991).
10. D.J. Wuebbles, "Weighing Functions for Ozone Depletion and Greenhouse Gas Effects on Climate," *Ann. Rev. Energy Environ.* **20,** 45-70 (1995).
11. World Meteorological Organization, *Scientific Assessment of Ozone Depletion* (1994); World Meteorological Organization, Global Ozone Research and Monitoring Project, Report No. 37D, Geneva (1995).
12. S. Solomon, J.B. Burkholder, A.R. Ravishankara, and R.R. Garcia, *J. Geophys. Res.* **99,** 20929-20935 (1994).
13. D.J. Wuebbles, P.S. Connell, and K.O. Patten, "Evaluating the Potential Effects of Halon Replacements on the Global Environment" in *Halon Replacements: Technology and Science,* A.W. Miziolek and W. Tsang, Eds., American Chemical Society, Washington, D.C. (1995).
14. S. Solomon, R.R. Garcia, and A.R. Ravishankara, *J. Geophys. Res.* **99,** 20491-20499 (1994).
15. D.J. Wuebbles, P.S. Connell, and K.O. Patten, "Evaluating the Potential Effects of Halon Replacements on the Global Environment" in *Halon Replacements: Technology and Science,* A.W. Miziolek and W. Tsang, Eds., American Chemical Society, Washington, D.C. (1995).
16. World Meteorological Organization, *Scientific Assessment of Ozone Depletion* (1994); World Meteorological Organization, Global Ozone Research and Monitoring Project, Report No. 37D, Geneva (1995).
17. T.J. Wallington, W.F. Schneider, J. Sehested, and O.J. Nielsen, "Hydrofluorocarbons and Stratospheric Ozone," *Faraday Discussions* **100,** 55 (1995).
18. World Meteorological Organization, *Scientific Assessment of Ozone Depletion* (1994); World Meteorological Organization, Global Ozone Research and Monitoring Project, Report No. 37D, Geneva (1995).
19. D.J. Wuebbles, *J. Geophys. Res.* **88,** 1433-1443 (1983).

20. World Meteorological Organization, *Scientific Assessment of Ozone Depletion, 1991,* World Meteorological Organization, Global Ozone Research and Monitoring Project—Report No. 25, Geneva (1991).
21. World Meteorological Organization, *Scientific Assessment of Ozone Depletion* (1994); World Meteorological Organization, Global Ozone Research and Monitoring Project, Report No. 37D, Geneva (1995).
22. D.J. Wuebbles, "Weighing Functions for Ozone Depletion and Greenhouse Gas Effects on Climate," *Ann. Rev. Energy Environ.* **20,** 45-70 (1995).
23. World Meteorological Organization, *Scientific Assessment of Ozone Depletion, 1991,* World Meteorological Organization, Global Ozone Research and Monitoring Project—Report, No. 25, Geneva (1991).
24. World Meteorological Organization, *Scientific Assessment of Ozone Depletion* (1994), World Meteorological Organization, Global Ozone Research and Monitoring Project, Report No. 37D, Geneva (1995).
25. S. Solomon, M.J. Mills, L.E. Heidt, and A.F. Tuck, *J. Geophys. Res.* **97,** 825-842 (1992).
26. T.J. Wallington, W.F. Schneider, J. Sehested, and O.J. Nielsen, "Hydrofluorocarbons and Stratospheric Ozone," *Faraday Discussions* **100,** 55 (1995).
27. P.S. Connell, D.E. Kinnison, D.J. Bergmann, K.O. Patten, D.J. Wuebbles, R.G. Caniel, C.K. Williamson, A.W. Miziolek, and R.E. Huie, "Environmental Effects of Halon Replacements: Considerations for Advanced Agents and the Ozone Depletion Potential of CF31," *Proceedings of the Halon Options Technical Working Conference,* Albuquerque, N. Mex., May 7-9 (1996).
28. World Meteorological Organization, *Scientific Assessment of Ozone Depletion* (1994); World Meteorological Organization, Global Ozone Research and Monitoring Project, Report No. 37D, Geneva (1995).
29. Intergovernmental Panel on Climate Change (IPCC), *Radiative Forcing of Climate Change,* Cambridge University Press, Cambridge, Great Britain (1995).
30. D.J. Wuebbles, "Weighing Functions for Ozone Depletion and Greenhouse Gas Effects on Climate," *Ann. Rev. Energy Environ.* **20,** 45-70 (1995).
31. A.A. Laths, D.J. Wuebbles, and J.A. Logan, *A. Geophys. Res.* **95,** 9971-9951 (1990).
32. Intergovernmental Panel on Climate Change (IPCC), *Radiative Forcing of Climate Change,* Cambridge University Press, Cambridge, Great Britain (1995).
33. J.S. Darnel, S. Solomon, and D.L. Albritton, *J. Geophys. Res.* **99** (1994).
34. See, for example, J.P. Friend, in *Scientific Assessment of Stratospheric Ozone,* Vol. II, Chapter X (1990).
35. T.J. Wallington, W.F. Schneider, J. Sehested, and O.J. Nielsen, "Hydrofluorocarbons and Stratospheric Ozone," *Faraday Discussions* **100,** 55 (1995).
36. J.A. Kaye, A.R. Douglass, C.H. Jackman, R.S. Stolarski, R. Sander, and G. Roland. "Two-Dimensional Calculation of Fluorine-Containing Reservoir Species," *J. Geophys. Res.* **96,** 12865-12881 (1991).
37. See Wallington et al., Chapter 3 in *Halon Replacements: Technology and Science,* ACS Symposium Series 611, American Chemical Society, Washington, D.C. (1995).
38. M.F. Kanakidou, F.J. Dentener, and P.J. Crutzen, "A Global Three-Dimensional Study of the Fate of HCFCs and HFC-134a in the Troposphere," *J. Geophys. Res.* **100,** 18781-18801 (1995).
39. T.K. Tromp, M.K.W. Ko, J.M. Rodriguez, and N.D. Sze, "Potential Accumulation of a CFC Replacement Degradation Product in Seasonal Wetlands," Nature **376,** 327-330 (1995).
40. World Meteorological Organization, *Scientific Assessment of Stratospheric Ozone: 1989,* Vol. II, WMO Global Ozone Research and Monitoring Project Report No. 20, Geneva (1990).
41. R.A. Cox and R. Lesclaux, "Degradation Mechanisms of Selected Hydrochlorofluorocarbons in the Atmosphere: An Assessment of Current Knowledge," in *Scientific Assessment of Stratospheric Ozone: 1989,* Vol. II, World Meteorological Organization, Global Ozone Research and Monitoring Project, Report No. 20, Geneva, (1990).
42. T.J. Wallington, W.F. Schneider, D.R. Worsnop, O.J. Nielsen, J. Sehested, W.J. Debruyn, and J.A. Shorter, "Environmental Impact of CFC Replacements—HFCs and HCFCs," *Environ. Sci. Technol.* **28,** 320A-326A (1994).
43. T.J. Wallington, M.D. Hurley, J.C. Ball, and E.W. Kaiser, "Atmospheric Chemistry of Hydrofluorocarbon 134a: Fate of Alkoxy Radical CF_3FHO, *Environ. Sci. Technol.* **26,** 1318-1324 (1992).
44. E.C. Tuazon and R. Atkinson, "Tropospheric Degradation Products of CH_2FCF_3 (HFC-134a), *J. Atmos. Chem.* **16,** 301-312 (1993).
45. T.J. Wallington, W.F. Schneider, D.R. Worsnop, O.J. Nielsen, J. Sehested, W.J. Debruyn, and J.A. Shorter, "Environmental Impact of CFC Replacements—HFCs and HCFCs," *Environ. Sci. Technol.* **28,** 320A-326A (1994).
46. R. Zellner, G. Bednarek, A. Hoffmann, J.P. Kohlmann, V. Meyers, and H. Saathoff, "Rate and Mechanism of the Atmospheric Degradation of 2H-heptafluoropropane (HFC-227)," *Ber. Bunsenges. Phys. Chem.* **98,** 141-146 (1994).

47. T.E. Mogelberg, J. Sehested, M. Bilde, T.J. Wallington, and O.J. Nielsen, "Atmospheric Chemistry of HFC-236fa: Spectrokinetic Investigation of the $CF_3CHO_2CF_3$ Radical, Its Reactions with NO, and the Fate of the CF_3CHOCF_3 Radical," *J. Phys. Chem.* **100,** 8882-8889 (1996).
48. T.E. Mogelberg, J. Platz, O.J. Nielsen, J. Sehested, and T.J. Wallington, "Atmospheric Chemistry of HFC-236fa: Spectrokinetic Investigation of the Radical $CR_3CHO_2CF_3$ Radical, Its Reactions with NO, and the Fate of the CF_3CHOCF_3 Radical," *J. Phys. Chem.* **99,** 5373-5378 (1995).
49. T.E. Mogelberg, J. Platz, O.J. Nielsen, J. Sehested, and T.J. Wallington, "Atmospheric Chemistry of HFC-236fa: Spectrokinetic Investigation of the Radical $CR_3CHO_2CF_3$ Radical, Its Reactions with NO, and the Fate of the CF_3CHOCF_3 Radical," *J. Phys. Chem.* **99,** 5373-5378 (1995).
50. A.R. Ravishankara, S. Solomon, A.A. Turnipseed, and R.F. Warren, "Atmospheric Lifetimes of Long-lived Halogenated Species," *Science* **259,** 194-199 (1993).
51. A.R. Ravishankara, S. Solomon, A.A. Turnipseed, and R.F. Warren, "Atmospheric Lifetimes of Long-lived Halogenated Species," *Science* **259,** 194-199 (1993).
52. Intergovernmental Panel on Climate Change (IPCC), *Radiative Forcing of Climate Change,* Cambridge University Press, Cambridge, Great Britain (1996).
53. Visscher et al., Nature **369,** 729 (1994).
54. AFEAS Workshop on the Environmental Fate of Trifluoroacetic Acid (1994).
55. A.R. Ravishankara, S. Solomon, A.A. Turnipseed, and R.F. Warren, "Atmospheric Lifetimes of Long-lived Halogenated Species," *Science* **259,** 194-199 (1993).
56. M.K. Ko, N.D. Sze, J.M. Rodriguez, D.K. Weisenstein, C.W. Heisy, R.P. Wayne, P. Biggs, C.E. Canosa-Mas, H.W. Sidebottom, and J. Treacy, "CF_3 Chemistry: Potential Implications for Stratospheric Ozone," *Geophys. Res. Lett.* 21, 101-104 (1994).
57. Review in Chapter 12 of World Meteorological Organization, *Scientific Assessment of Ozone Depletion* (1994).
58. T.J. Wallington, W.F. Schneider, J. Sehested, and O.J. Nielsen, "Hydrofluorocarbons and Stratospheric Ozone, *J. Chem. Soc.*, Faraday Discuss. **100,** 55-64 (1995).
59. T.J. Wallington, W.F. Schneider, J. Sehested, and O.J. Nielsen, "Hydrofluorocarbons and Stratospheric Ozone, *J. Chem. Soc.*, Faraday Discuss. **100,** 55-64 (1995).
60. AFEAS Workshop on the Environmental Fate of Trifluoroacetic Acid (1994).
61. T.J. Wallington, W.F. Schneider, D.R. Worsnop, O.J. Nielsen, J. Sehested, W.J. Debruyn, and J.A. Shorter, "Environmental Impact of CFC Replacements—HFCs and HCFCs," *Environ. Sci. Technol.* **28,** 320A-326A (1994).
62. T.J. Wallington, W.F. Schneider, D.R. Worsnop, O.J. Nielsen, J. Sehested, W.J. Debruyn, and J.A. Shorter, "Environmental Impact of CFC Replacements—HFCs and HCFCs," *Environ. Sci. Technol.* **28,** 320A-326A (1994).
63. T.K. Tromp, M.K.W. Ko, J.M. Rodriguez, and N.D. Sze, "Potential Accumulation of a CFC Replacement Degradation Product in Seasonal Wetlands," *Nature* **376,** 327-330 (1995).
64. Frank et al., *Nature* **382,** 34 (1996).

4

Navy-Specific Issues

Halons are recognized as ideal fire extinguishing agents, particularly for fighting fires caused by flammable liquids and explosive gases. They are highly effective in extinguishing fires in minimal time. They are non-corrosive and, when deployed at the recommended volume densities, they are non-toxic. Because of these characteristics, halon 1301 (a total flooding agent) and halon 1211 (a streaming agent) are widely used on board Navy ships, in aircraft, and at shore facilities, as they arc throughout the civil sector.

Unfortunately halons do contribute to depletion of Earth's ozone layer, a peril recognized by the 1987 Montreal Protocol. Even though lower in emissions than their refrigerant cousins, the greater ozone-depleting characteristics of halons per pound have resulted in a mandated cessation of production in the United States, some 25 years after halons were introduced in the Navy as fire extinguishing agents.

Under an executive order effective in January 1994, halons can no longer be manufactured in the United States. But because of the difficulty of finding a suitable fire extinguishing substitute for halons 1301 and 1211, the military services are permitted to use these chemicals for mission-critical purposes, such as fire fighting, in existing platforms (ships, aircraft, weapons, vehicles) until the current halon supply or "bank" is exhausted. Each service is expected to live within its own halon budget, and transfers between service accounts are permitted only with approval of the Secretary of Defense. The hope, then, is that (1) the Navy's halon supply is sufficient to protect existing platforms until they are retired from service or scrapped, and (2) there is sufficient time to develop and test suitable replacement fire extinguishing agents and dispensing systems for next-generation platforms.

In its deliberations, the committee recognized the importance of assessing the potential and need for finding a drop-in halon replacement, given the challenge faced by the Navy to reduce the risk, or perception of risk, to combat readiness and peacetime safety that might flow from either early rescission of authority to use halon—a possible result of increasing international pressure to stop any use of halon—or the exhaustion of halon supplies that could result from now unforeseen demands of a prolonged, major war on the scale of World War II, Korea, or Vietnam.

In assessing the potential for finding a suitable drop-in, the committee began with the postulate that any replacement agent or system must possess at least five attributes: (1) performance that meets the fire extinguishing requirement, (2) low toxicity, (3) acceptable environmental properties, (4) size and weight that can be accommodated on existing platforms, and (5) procurement and installation that are not prohibitively expensive. The committee accepted the premise that such a replacement, if identified, would very likely not match the exceptional fire extinguishing performance of halon on a space and weight basis, but also that it need not do so. What matters more is adequacy of performance, feasibility of installation, and affordability.

Determining the need for a drop-in agent involved assessing (1) the probability that halon systems now installed may have to be replaced and (2) the effectiveness of non-halon systems the Navy has selected for its next-generation ships and aircraft as well as the feasibility of installing them in current platforms. In addressing these issues, the committee sought answers to the following key questions: Is the Navy's supply of halon sufficient to last until the current classes of ships and aircraft are retired from service? Are the halon replacement systems selected by the Navy for new-design ships and aircraft (HFC-227ea, water mist, HFC-125, gas generator) adequate in performance and environmentally satisfactory? Is retrofit of HFC-227ea/water mist (ships) and HFC-125/gas generators (aircraft) into existing platforms technically feasible? What is the estimated cost of retrofit? Is a drop-in replacement for halon very near at hand, a scientific possibility in some reasonable time if sufficient resources are applied, or an illusory target unworthy of investing R&D funds?

This chapter summarizes current halon 1301 ship and aircraft installations and the Navy's investment in halon systems, discusses systems planned for new-design ships and aircraft, assesses the current and projected status of the Navy's halon inventory, and finally, examines the potential for retrofit of non-halon systems into existing platform designs. The chapter closes with the committee's findings and recommendation on Navy-specific management issues. Halon systems in shore facilities were not considered by the committee since these do not fall into the mission-critical category and are to be replaced by the year 2000. Appendix A takes a brief historical look at the introduction of halon fire extinguishing systems in the Navy and discusses in some detail the Navy's need for and use of extinguishing systems currently installed aboard ships and aircraft.

SHIP SYSTEMS

Summary of Halon 1301 Ship Installations

Table 4.1 lists all halon 1301-equipped ships in commission today as well as those current designs with units yet to be built. It can be seen that some 1.4 million pounds of halon are now installed, with 441,000 pounds scheduled for commitment to new-construction vessels in the future.

A review of the decommissioning schedule shows that few ships will be removed from active service for the next 15 to 20 years. Beginning in 2015, however, the pace of decommissioning picks up and, by 2025, over 50% of the currently installed halon systems will have been taken out of service, with the largest contributors to this halon reduction being LPD-4 class amphibious ships, FFG-7 class frigates, DD-963 destroyers, and the remaining fossil-fueled aircraft carriers.

Looking at the halon 1301-equipped ships yet to be built, the DDG-51 class Aegis destroyers appear to warrant special attention. Thirty-seven of these vessels are scheduled to be built in the future, with *construction* phased by blocks or "flights" in which accumulated design changes are incorporated. The 37 ships will each require 8995 pounds of agent for a total of 332,815 pounds of installed halon 1301.

Not shown in Table 4.1 are. ships of the Military Sealift Command (MSC). As of August 1996, MSC operated 46 ships incorporating 331 halon 1301 systems containing 509,000 pounds of agent. The committee understands that MSC vessels fall outside the normal purview of the Office of Naval Research and the Naval Sea Systems Command, and therefore, details of installations and MSC's plans for the future were not considered during the course of the study.

Investment in Ship Halon Systems

The committee inquired into the investment the Navy has made to date in halon 1301 installations in the current fleet. Taking the mix of ship sizes in the various classes, NavSea engineers calculated that the average cost for an individual system in today's dollars is about $300,000. Installation of piping, bottles, and control equipment accounts for $275,000, with the remaining $25,000 being the cost of hardware and agent. Multiplying this average cost by the total number of systems listed in Table 4.1 yields an investment of $665 million. And if new-construction plans are executed as now planned and shown in the table, this figure will grow by another $166 million to an investment totaling $831 million. To make the investment picture complete, the cost of RDT&E would have to be added, but this information was not available to the committee.

Fire Extinguishing Systems for New-Design Ships

The Navy is shifting to non-halon systems in its next-generation, new-design ships. These include the LPD-17 amphibious ship class, the next aircraft carrier (CVN-76), and a proposed new surface combatant class (SC-21) that will follow the DDG-51 production run.

Table 4.1 U.S. Navy Shipboard Halon 1301 Systems

Ship Class	Current Number of Ships per Class	No. of Ships in Class Under Construction or Planned	Number of Systems per Ship	Amount of Halon per Ship (lb)	If Banked, Quantity of Two-Shot Systems per Ship	Systems per Class	Amount of Halon per Class (lb)	No. of Systems in Construction	Amount of Halon in Construction (lb)	Banked (B) or Modular (M) Systems	Estimated Decom. Year for Last Ship of Class	Assumed Service Life of Ship Class (years)
AO-177	5		13	13630	3	65	68150			B	2023	40
AOE-1	4		5	10125		20	40500			M	2010	40
AOE-6	3	1	13	24820	10	39	74460	13	24,820	B	2038	40
ARS-50	4		6	2375	2	24	9500			B	2026	40
AS-39	1		3	8360		3	8360			M	2021	40
CG-47	27		8	4905	1	216	132435			B	2034	40
CV-59/63	4		14	20500		56	82000			M	2018	50
CVN-65	1		7	4245		7	4245			M	2013	N/A
CVN-68	5		7	3500		35	17500			M	2039	50
CVN-73	2	1	14	4340	0	28	8680	14	4,340	B	2048	50
DD-963	31		13	6275		403	194525			M	2023	50
DDG-993	4		6	5000		24	20000			M	2022	40
DDG-51	16	37	12	8995	9	192	143920	444	332,815	B	2047	40
FFG-7	45		15	5410	12	675	243450			B	2018	40
LCC-19	2		6	6625		12	13250			M	2010	N/A
LHA-1	5		8	11375		40	56875			M	2020	40
LHD-1	4	3	14	18220	3	56	72880	42	54,660	B	2040	40
LPD-4	13		5	2975		65	38675			M	2011	40
LSD-36	5		5	4005		25	20025			M	2012	40
LSD-41	11	1	10	10285	4	110	113135	10	10,285	B	2038	40
LST-1179	2		10	4525		20	9050			M	2011	40
MCM-1	14		3	1810	2	42	25340			B	2029	40
MCS-12	1		5	6540		5	6540			M	2010	40
MHC-51	6	6	5	2375	4	30	14250	30	14,250	B	2034	35
PC-1	13		2	1000	2	26	13000			B	2024	30
TOTALS	228	49				2218	1430745	553	441,170			

In choosing a fire extinguishing approach for these new ship designs, the Navy evaluated a wide range of new technologies to replace halon 1301. After extensive testing, heptafluoropropane (HFC-227ea), a commercially available gaseous agent, and a Navy-designed high-pressure water mist system were chosen. Either system has fire extinguishing performance at least equivalent to that of halon 1301.

In the LPD-17, water mist will be used in the machinery spaces and HFC-227ea elsewhere, with the former chosen because of lower cost. The CVN will employ HFC-227ea solely, since no requirement exists for main machinery space systems. The SC-21 ship design is still in the concept stage, and decisions about fire extinguishing systems have yet to be made.

HFC-227ea

The HFC-227ea total flooding gas system, largely using existing 600-psi halon 1301 hardware, is to be installed in flammable liquid storage and issue rooms as well as other small segregated compartments in both the LPD-17 and CVN-76. These systems are relatively small in size (<< 100 lb of agent) and function identically to today's halon 1301 systems.

Water Mist

Water mist technology has recently been the subject of considerable interest because it offers lower water demand (than standard water sprinkler systems) and the promise of the ability to extinguish fires in obstructed spaces (a key feature of halons and other gaseous agents). Advantages include low agent cost, absence of toxicity and environmental problems, effectiveness in suppressing flammable liquid pool and spray fires, and potential for explosion suppression. Details of developments in the field are presented in Appendix D. Water mist systems employ high pressures (circa 1000 psi) and nozzles designed to produce drops distributed about the 100-micron size range. Drops smaller than 50 microns in diameter begin to exhibit characteristics of a gas, resulting in lower fallout losses and to some extent, the ability to diffuse around obstructions.

The Navy has subjected commercial and Navy-designed water mist systems to extensive testing in the *ex-USS Shadwell* facility. As a result, the Navy-designed system has been chosen for use in all machinery spaces in the LPD-17, an amphibious ship class of new design. The system is driven by two independent 250-hp pumps, supplying water at 1000 psi to nozzles in each space.

The arrangement of the machinery spaces in LPD-17 makes for a particularly efficient system. Redundant pumps are provided, forward and aft, and valving arrangements direct water to any space using either of the two pumps. Nozzles are installed with approximately 100-sq.-ft spacing, with water flow in the range of 2- to 3-gpm per nozzle. Nozzles are positioned in a uniform grid pattern in the overhead of each space and at the intermediate deck level.

The high water flow demand of the system, that is a disadvantage of the Navy design, precluded the use of pressurized water cylinders because of the significant space, cost, and weight impact. Hence, pumps were the only feasible option. Each pump motor is in the range of 200 hp, supplying 225 gpm at 1000 psi. Modem electrical power distribution systems in these new ships incorporate improved survivability features, and the Navy feels confident that a reliable power source will be available for the pumps even under battle damage conditions.

While the water mist system is effective, has no adverse environmental impact, and makes economic sense in the large space application, it is at present less suited for use in small spaces spread about a ship. In such instances, a water mist system may be more expensive and heavier than an HFC-227ea system because of the need for piping and pump redundancy, just the reverse of the large, concentrated machinery space application. To address this shortcoming, the Navy has funded an R&D program to evaluate a water mist system for small spaces which employs pressurized water containers rather than a pump system.

Table 4.2 Naval Aircraft Fire Extinguishing Systems—Current Status of Halon Applications

AIRCRAFT TYPE	YEAR OUT OF SERVICE[a]	NUMBER OF AIRCRAFT[b]		APPLICATION				WEIGHT (lb)				HALON Total Pounds for Aircraft Type	REMARKS
		Current	2028	Engine	Dry Bay	Portable[c]	Other	Halon[d]	Bottle[e]	Plumbing[f]	System	Platform	
Fixed Wing													
P-3	>2028	351	42	X		X	X-APU	46	53	13	111	15,970	
F/A-18	>2028	826	6	X				6	9	10	23	4,045	
A-6E	1997	85		X			X-FUEL	9	69	--	--	7,225	Retired 1997
F-14	2028	422	1	X			X-OWAC	16	26	14	55	6,465	
E-2C	2022	123		X		X		13	17	4	34	1,600	
EA-6B	>2028	56	5	X				30	30	4	64	1,680	
C-2A	2020	38		X		X	X-APU	16	25	4	44	590	
S-3	>2028	134	6			X	X-APU	1	9	1	11	153	
E-6A	>2028	16	16	X		X	X-APU	35	72	3	109	550	
C-130	2019	109		X		X		54	46	28	128	5,890	
T-44A	2000	45		X		X		5	15	3	23	225	
F/A-18E/F	2010	7		X				6	12	9	27	395	EMD only
C-9	>2028	29	26	X		X[g]		17	32	5	53	490	
C-12	>2028	80	22	X		X		5	15	3	23	840	
C-20	>2028	7	7	X			X-APU	14	27	2	42	96	
T-39	>2028	14	4	X				13	17	3	33	180	
Rotary Wing													
H-53E	>2028	199	30	X		X	X-APU	18	50	6	74	3,580	
SH-60	>2028	369	66	X		X		5	20	3	28	1,240	
H-53A/D	2024	128		X		X	X-APU	14	37	5	55	1,730	
CH-46	>2028	253	12	X		X		6	19	5	30	1,520	
SH-2	>2028	89	10	X		X		5	20	3	28	445	
H-3	2012	100		X		X		5	20	3	28	500	
AH-1	>2028	213	47	X		X		4	21	3	28	640	Current upgrade
UH-1	>2028	150	28	X		X		4	21	3	28	600	Current upgrade
V-22	2007	2		X		X		5	20	3	28	10	FSD only
											Grand Total	56,659	

[a] Estimated based on life extension programs.
[b] Rough estimates, not exact figures.
[c] Except for E-2C aircraft, it is planned that in 1997 all platforms using halon portables are to be retrofitted with CO_2 portables.
[d] Excludes portable halon weight (2.75 lbs per extinguisher).
[e] Empty bottle weights derived from MIL-C-22284. Excludes empty portable bottle weight.
[f] Estimated based on weight per foot of 0.5-inch diameter stainless-steel tubing. Excludes valves, fittings, switches, gauges, and mounting hardware.
[g] Each C-9 has two halon 1211 portables.

AIRCRAFT SYSTEMS

Summary of Halon 1301 Aircraft Installations

Table 4.2 lists all naval aircraft that have halon 1301 fire extinguishing systems installed, indicating the type and number of aircraft, the year the model goes out of service, halon application, and system weight. With the exception of the S-3, all multiengine planes are equipped with engine bay/nacelle extinguishers and all cabin aircraft carry portable bottles for hand-held use. Planes with auxiliary power units are protected, and the F- 14 also has a system covering an area above the wing, aft of the cockpit, which has proven to be vulnerable to hydraulic fluid fires. The retirement of the A-6E attack plane in 1997 will remove the only halon inerting system in the naval aviation inventory. No existing naval aircraft employs halon for dry bay fire extinguishing purposes.

Significantly, the total quantity of halon installed in naval aircraft is a small fraction (5%) of that in ships—65,000 lb versus 1.4 million lb. Proportionally, however, the aircraft contribution to annual halon releases is far greater than that of ships, constituting some 40% of the total. The Navy is working actively to reduce these releases of halon from aircraft and has succeeded in lowering the annual release rate by over one-half in the last 4 years.

Investment in Aircraft Halon 1301 Systems

As with ships, the committee was interested in determining the investment the Navy has made in halon 1301 aircraft systems. Upon request, engineers oft he Naval Air Systems Command analyzed one aircraft type in each of four size categories—very small, small, medium, and large, similar to the SH-60, F-18C/D, P-3, and C-130, respectively. As in the ship case, the halon 1301 system cost for an individual aircraft was determined by adding installation, hardware, and agent cost elements. This figure was, in turn, multiplied by the number of aircraft in the respective size category to obtain a category subtotal. Combining the four categories resulted in a total amount, in 1996 dollars, of $140 million. While this figure is based on a somewhat cursory analysis, it nevertheless gave the committee a rough approximation of the halon 1301 investment in the Navy' s current fleet of aircraft. To get a complete investment picture, the cost of installing halon systems in aircraft still being manufactured (e.g., F-18C/D) would, of course, have to be added as well the cost of initial system RDT&E; however, this information was not available to the committee.

Fire Extinguishing Systems for New-Design Aircraft

The Navy is shifting to non-halon fire extinguishing systems for its next-generation, new-design aircraft, and the proposed Joint Strike Fighter (JSF) will use an alternative system as well. Two approaches have been selected and tested for use in the next naval aircraft, the F-18E/F and V-22. The JSF will not enter engineering and manufacturing development until the turn of the century; hence, the fire extinguishing needs and technical approach for this joint services aircraft have not yet been selected.

HFC-125

The V-22 will employ both new approaches—an HFC-125 compressed gas system and one based on new inert gas generator technology. HFC-125 liquefied gas, the 1301 replacement selected by the three military services, will be used in the engine bays. These HFC-125 systems are identical in architecture and function to current halon 1301 systems, except that three times the weight of agent is required to meet the fire extinguishing requirement. Gas generators will be employed elsewhere in the V-22 and in both engine bays and dry bay areas of the F-18E/F. Gas generator technology and its specific application in these two new aircraft are discussed below.

Inert Gas Generator Technology

Inert gas generator technology for aviation fire extinguishing applications has been developed almost entirely by the Naval Air Systems Command. While the underlying propellant technology is well understood, extension to fire suppression applications has posed significant development and engineering challenges.

The technical basis for the suppression of diffusion or premixed hydrocarbon/air flames with inert gases is well established. Extinction of flames will occur at a specific concentration of an inert gas in air. There are differences in extinguishing concentration with fuel type, atmospheric pressure, and oxygen concentration, but the basic principle has been well established. However, design of a propellant system that will produce adequate quantities of inert gas quickly enough and distribute the gas to all locations in a high-air-flow environment is a substantial challenge.

An analogous technology, combustion-generated aerosols or pyrotechnically generated aerosols, has been used as a halon replacement. In these devices, a combination of solid propellant and binder produces a mixture of inert gases and fine solid particulate. The solid particulate is, in principle, a more efficient fire suppressant than are inert gases only. These technologies have been pursued overseas following initial development in the USSR.

The inert gas generator system developed and tested by the Navy is designed to retain most of the solids inside the generator and to minimize the discharge of particulate into the engine or dry bay. This helps to resolve the issue of collateral damage associated with combustion-generated aerosols and simplifies agent mixing and the distribution problem.

The performance requirements for the two aviation uses (engine bay and dry bay) of inert gas generators are substantially different. The engine bay threat is a liquid fuel diffusion flame in a complex flow geometry with high air flow rates. The system must discharge within a few seconds and produce sufficient agent to extinguish the diffusion flame at any location in the engine nacelle. Challenges include a widely variable air flow, numerous flame stabilization points, and a highly obstructed flow geometry.

Protection of aviation dry bays requires suppression of an incipient premixed liquid/fuel aerosol in the presence of a hot ignition source. This is in effect an explosion or deflagration suppression problem. Here, the requirement is for detection and suppression of the explosion kernel within of tens of milliseconds. The agent must be produced and distributed throughout the protected volume or directed locally around the explosion kernel. Sufficient duration of agent flow must be provided to prevent reignition of the fuel or flashback from a remote unextinguished flame.

Inert gas generators rely on the production of CO_2 or nitrogen at high rates through the combustion of solid propellants. As shown in Figure 4.1, the hardware and process are analogous to sodium azide air bag inflators in automobiles.

F/A-18E/F Engine Nacelle/Dry Bay

The inert gas generator system developed for the F/A-18E/F engine bay and tested in real-scale live fire testing demonstrated equivalent or lower space and weight requirements relative to one using halon 1301. Although a thin film of particulate is developed within the engine bay, its impact has been determined to be inconsequential.

Inert gas generator systems were also developed for F/A-18E/F dry bay deflagration suppression. These systems use multiple (6 to 10) gas generators, actuated by 1 of 14 optical (flame) fire detectors. The gas generators are fired in either a fixed sequence or in a sequence determined by which of the 14 optical detectors is first triggered. A typical system employing gas generators is similar to a halon 1301 installation for extinguishing engine fires (see Figure A.3, Appendix A) except that the generator replaces the halon bottle.

The configuration currently being tested consists of six 157-gram gas generators, 14 optical fire detectors, and a fixed generator firing sequence. These tests are being performed on an F-18C test platform.

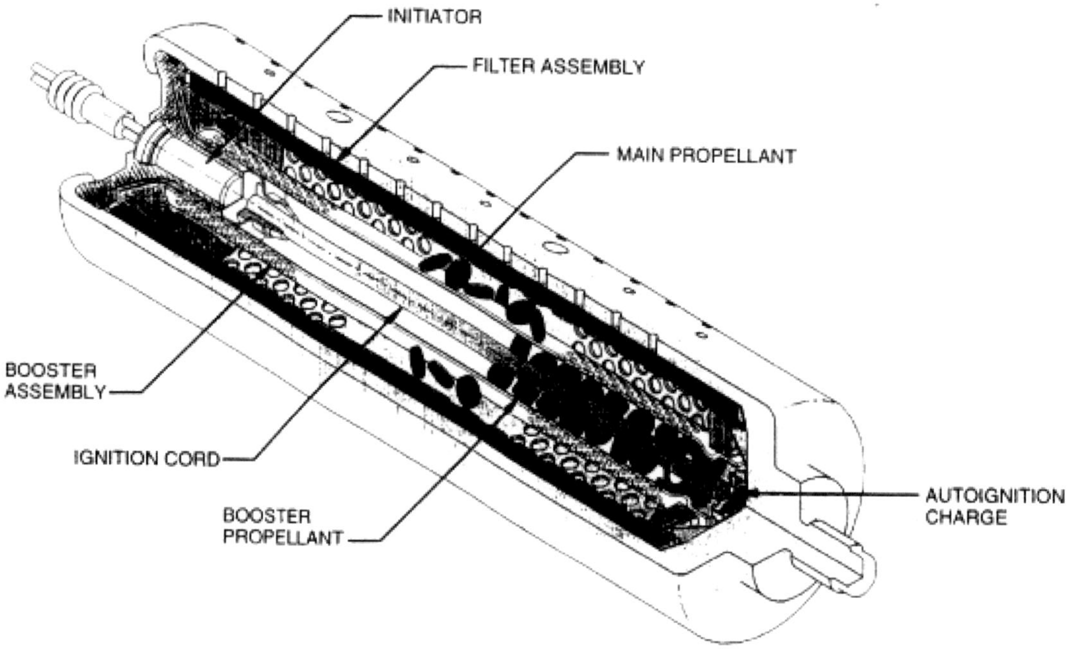

Figure 4.1
Typical azide inflator used in automotive industry; not for use in aircraft.

V-22 Midwing Protection

Inert gas generators are also being used to protect the midwing and wing bay volume of the V-22. This area of the aircraft must be protected as one single volume due to the multiple flow paths for both fire suppression agent and fuel/flame. The design of inert gas systems for the midwing required optimization of propellant loading for fire extinguishment with simultaneous minimization of the threat of overpressurization due to excessive gas production. In addition, a generator firing controller and associated logic were developed to actuate the needed generators in the proper sequence to locally extinguish the fire and to maintain an inert atmosphere long enough to prevent reignition or flashback.

The final design consists of 17 gas generators of five different sizes and 16 optical detectors with the requisite generator sequencing and detection logic. This design has been subjected to actual-scale live fire testing and was successfully qualified.

Inert Gas Generator Development Summary

Inert gas generators for naval aviation applications have been shown to be an effective replacement for halon 1301. They have demonstrated performance similar to that of halon 1301 in high-challenge engine nacelle fires, with space and weight requirements lower than those for halon 1301. The rapid development, qualification, and deployment of this technology into the F/A-18E/F and V-22 is a remarkable achievement, accomplished almost wholly within the naval aviation community, and is certainly worthy of special note.

Given the flexibility of the design and the low space and weight requirements, the gas generator may be a suitable candidate for retrofit replacement of halon 1301 systems in selected existing aircraft should that become necessary in the future. However, an impediment to arriving at a decision to retrofit such a system or to incorporate gas generators into new designs is the insufficiency of engineering tools for evaluating gas generator performance, notably in the flow, mixing, and flame extinction processes, thereby causing undue reliance on expensive and time-consuming full-scale tests. Adaptation of existing fluid flow and combustion models to inert gas generator systems would greatly facilitate future design efforts, with potential significant cost and time savings resulting from a reduction in requirements for full-scale testing.

NAVY HALON INVENTORY

The committee inquired into the amount of halon installed in ships and aircraft, its predicted usage over the various platform life spans, and the status of non-installed halon reserves available to the Navy, sometimes referred to as the "bank" or "stockpile." By direction of the Office of the Secretary of Defense, halon reserves are under the central control of the Defense Logistics Agency. However, each military service has been allocated a portion of this reserve in accordance with its predicted needs at the time the Executive Order was issued banning halon production in the United States.

The Navy manages its halon usage, recycling programs from ships and aircraft going out of service, and withdrawals from the reserve. Like the other services, the Navy is expected to meet future needs for halon 1301 by recycling, reducing releases caused by human error, and emphasizing a lower-release approach in training. Shortfalls are made up by drawing from the reserve allocation, within which the Navy is expected to remain unless authorized to exceed the limit.

Table 4.3 contains projections of halon use, anticipated recovery of halon from retiring ships and aircraft, and the resultant impact on the Navy halon reserve out to the end of the service lives of the various halon-equipped platforms. While ship and aircraft counts can be expected to change somewhat from year to year, and usage and recovery data revised as recovery experience is gained and data collection methods improved, the presentation nevertheless highlights several points of note.

First, aircraft account for only a minute portion of the installed base of halon while contributing 40% of the releases. Second, as Figure 4.2 shows graphically, the reserve is projected to be depleted gradually over the years through releases and new installations until reaching a low point in the 2030 time period. After this, the reserve gradually builds due to recycling and lower annual releases occasioned, in turn, by there being fewer halon-equipped ships in service.

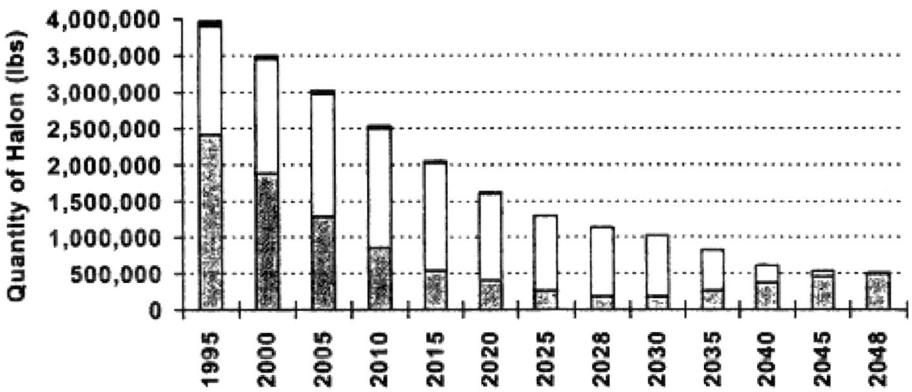

Figure 4.2
Navy total halon 1301 reserve bottom shaded portion of bars) including Military Sealift Command (MSC) reserve quantities and net usage. Note, however, that number of MSC ships (46 ships currently) and installed charges (approximately 0.5 million lb currently) are not included. The installed base on ships is indicated in the middle portion of each bar, and the installed base on aircraft at the top of each bar.

Table 4.3 FY 96 Navy Halon 1301 Projections and Navy Reserve

	Ships							
FY	Average Installed Base (lb)	Average Number of Ships	New Ship Construction	New Construction Usage (lb)	Number of Ships Retired	Operational Usage (4.45%) (lb)	Chemical Recovered (lb)	Net Amount Required (lb)
1995	1,484,430	239	18	153,460	19	65,463	80,213	138,710
1996	1,483,688	237	12	75,640	15	65,431	81,197	59,874
1997	1,483,150	237	1	58,950	3	65,407	13,769	110,588
1998	1,517,018	239	8	96,020	9	66,900	47,296	115,625
1999	1,538,170	239	4	29,360	3	67,833	10,857	86,336
2000	1,562,528	240	4	45,205	2	68,907	7,238	106,874
2001	1,597,950	243	4	35,980	0	70,470	0	106,450
2002	1,629,433	247	3	26,985	0	71,858	0	98,843
2003	1,656,418	250	3	26,985	0	73,048	0	100,033
2004	1,678,340	252	3	26,985	1	74,015	7,088	93,912
2005	1,694,940	253	3	26,985	3	74,747	7,452	94,280
2006	1,712,530	254	3	26,985	2	75,523	5,702	96,806
2007	1,719,650	253	3	26,985	6	75,837	22,110	80,712
2008	1,702,938	248	0	0	6	75,100	20,178	54,922
2009	1,679,973	244	0	0	3	74,087	11,974	62,113
2010	1,644,125	236	0	0	12	72,506	38,213	34,293
2011	1,583,583	227	0	0	7	69,836	46,547	23,289
2012	1,541,160	221	0	0	4	67,965	12,845	55,120
2013	1,522,108	217	0	0	4	67,125	13,829	53,296
2014	1,504,475	214	0	0	3	66,347	10,857	55,490
2015	1,485,828	210	0	0	4	65,525	15,250	50,275
2016	1,448,943	204	0	0	8	63,898	36,390	27,509
2017	1,393,820	196	0	0	9	61,467	40,782	20,685
2018	1,315,548	185	0	0	13	58,016	68,800	(10,784)
2019	1,238,755	174	0	0	8	54,629	38,710	15,919
2020	1,185,550	167	0	0	7	52,283	35,777	16,506
2021	1,132,050	160	0	0	6	49,923	39,123	10,800
2022	1,093,885	156	0	0	3	48,240	14,308	33,932
2023	1,068,680	150	0	0	8	47,129	20,979	26,150
2024	1,045,043	141	0	0	10	46,086	12,114	33,973
2025	1,021,405	131	0	0	10	45,044	20,979	24,065
2026	991,105	123	0	0	7	43,708	21,441	22,267
2027	958,543	115	0	0	8	42,272	24,147	18,125
2028	933,675	109	0	0	5	41,175	10,668	30,507
2029	888,395	102	0	0	8	39,178	52,724	(13,546)
2030	834,718	95	0	0	6	36,811	22,425	14,387
2031	794,500	88	0	0	9	35,037	33,880	1,157
2032	738,880	78	0	0	10	32,585	43,988	(11,403)
2033	700,103	72	0	0	3	30,875	10,301	20,574
2034	662,005	67	0	0	7	29,194	43,036	(13,842)
2035	560,065	58	0	0	11	24,699	99,680	(74,981)
2036	461,235	49	0	0	6	20,340	38,682	(18,342)
2037	404,755	43	0	0	6	17,850	40,390	(22,540)
2038	331,253	37	0	0	7	14,608	62,514	(47,905)
2039	271,358	31	0	0	4	11,967	21,340	(9,373)
2040	233,513	27	0	0	4	10,298	31,644	(21,346)
2041	192,920	23	0	0	4	8,508	25,186	(16,678)
2042	159,268	19	0	0	4	7,024	21,928	(14,904)
2043	130,113	16	0	0	3	5,738	18,890	(13,152)
2044	103,128	13	0	0	3	4,548	18,890	(14,342)
2045	76,143	10	0	0	3	3,358	18,890	(15,532)
2046	46,988	6	0	0	4	2,072	21,928	(19,855)
2047	17,833	3	0	0	3	786	18,890	(18,103)
2048	2,170	1	0	0	1	96	3,038	(2,942)

Table 4.3

FY	Aircraft								MSC Projected Net Usage (lb)	End of FY Remaining Navy Stockpile (lb)
	Average Installed Base (lb)	Average Number of Aircraft	New Aircraft Construction	New Construction Usage (lb)	Number of Aircraft Retired	Operational Usage (lb)	Chemical Recovered (lb)	Net Amount Required (lb)		
1995	69,472	3,669	88	1,090	100	48,290	8,526	40,854	136	2,412,051
1996	65,946	3,636	41	354	96	45,769	5,985	40,138	136	2,437,692
1997	61,904	3,576	40	334	105	43,181	5,772	37,743	136	2,289,225
1998	59,428	3,528	31	288	61	40,345	5,757	34,876	136	2,138,588
1999	52,986	3,498	29	267	60	39,891	3,004	37,154	136	2,014,962
2000	46,646	3,458	24	252	73	26,666	2,952	23,966	136	1,883,985
2001	45,816	3,369	7	74	82	26,338	2,789	23,623	136	1,753,777
2002	44,862	3,325	10	105	76	25,845	3,022	22,925	136	1,631,873
2003	43,804	3,245			94	25,355	2,886	22,469	136	1,509,235
2004	42,572	3,155			87	24,709	3,120	21,589	136	1,393,597
2005	41,267	3,058			106	24,133	3,064	21,069	136	1,278,112
2006	39,905	2,960			90	23,489	3,192	20,297		1,161,009
2007	38,417	2,875			116	22,898	3,023	19,875		1,060,422
2008	36,890	2,749			100	22,055	3,340	18,715		986,785
2009	35,486	2,645			108	21,405	2,905	18,500		906,172
2010	34,024	2,540			102	20,735	2,890	17,845		854,034
2011	32,560	2,442			95	20,009	2,945	17,064		813,680
2012	31,160	2,341			106	19,347	2,846	16,501		742,059
2013	29,706	2,235			106	18,686	2,747	15,939		672,824
2014	28,324	2,139			86	17,984	2,834	15,150		602,183
2015	26,938	2,042			108	17,391	2,495	14,896		537,012
2016	25,218	1,922			133	16,479	2,790	13,689		495,814
2017	23,106	1,759			193	15,519	3,055	12,464		462,664
2018	20,904	1,585			154	14,042	3,483	10,559		462,889
2019	18,174	1,421			174	12,701	3,083	9,618		437,357
2020	15,569	1,245			178	10,756	4,274	6,482		414,364
2021	13,626	1,073			167	9,656	2,688	6,968		396,596
2022	11,746	922			134	8,299	2,820	5,479		357,185
2023	10,189	788			134	7,198	2,338	4,860		326,175
2024	8,656	661			120	6,101	2,054	4,047		288,155
2025	7,204	550			103	5,058	2,099	2,959		261,131
2026	6,032	450			97	4,199	1,670	2,529		236,335
2027	5,061	363			76	3,398	1,477	1,921		216,289
2028	3,754	277			97	2,633	1,216	1,417		184,365
2029										197,911
2030										183,524
2031										182,367
2032										193,770
2033										173,196
2034										187,038
2035										262,019
2036										280,361
2037										302,901
2038										350,806
2039										360,179
2040										381,524
2041										398,203
2042										413,106
2043										426,258
2044										440,599
2045										456,131
2046										475,986
2047										494,090
2048										497,032

NOTES FOR TABLE 4.3:

1. NAVAIR inventory, installed charge, and requirements data are from the beginning of FY 1995 through the end of FY 2028.
2. The hull and aircraft numbers are the average numbers in service throughout the respective fiscal years. This applies to the installed charge values as well.
3. Analysis includes projected net usage of 136 lb per year for MSC from FY 1995 through FY 2005.
4. Stockpile balances on September 30, 1995, and July 1, 1996, are the starting balance on January 1, 1995, net of the total Defense Supply Center and Cheatham Annex issues and returns to DSC, to date. Beyond July 1, 1996, the stockpile balance is estimated using the projected usage and recovery for each fiscal year.
5. The NAVSEA recovery rate is assumed to be 70%.
6. NAVSEA hulls and installed charges do not include any landing craft air cushions.

Finally, the Navy has about $29 million invested in the halon 1301 reserve based on a recent large-purchase price of $12/lb. At the current market price for small lots, the value of this reserve could be as high as $85 million. It is anticipated that the price will rise as time passes and as the impact of manufacturing cessation is felt worldwide.

Assuming that the predictions are accurate, it appears that the Navy has sufficient agent in hand to support halon-equipped ships and aircraft until they go out of service. This conclusion is valid so long as the United States is not involved in a major war, there is no rescission of the current authorization to use halon for military-critical purposes, and the Navy's inventory is well managed. Given the minimal reserve forecast at the 2030 stockpile nadir, however, the Navy may wish to consider adding modestly to its reserve in the near term as a hedge against uncertainty or, alternatively, electing to install non-halon fire extinguishing systems in selected new-construction vessels such as the DDG-51.

RETROFITTING NON-HALON SYSTEMS IN EXISTING SHIPS AND AIRCRAFT

When it came to a consensus that a "no-penalty" drop-in substitute for halon 1301 was not anywhere near at hand, the committee thought it prudent to explore the technical feasibility and cost of retrofitting, in existing platforms, the fire extinguishing system approaches already selected for new-design ships and aircraft. The committee thought it important to make such a determination as a hedge against the possibility, however unlikely, that use of halon might be proscribed before existing ships and aircraft were retired or that the reserves might prove to be insufficient because of mismanagement or future unanticipated high usage. The systems considered were HFC-227ea and water mist for ships, and HFC-125 and inert gas generators for aircraft, all of which have been described previously.

Ship Retrofit

HFC-227ea Systems

Since more than twice the weight and storage volume of HFC-227ea, relative to halon 1301, is required to achieve adequate extinguishing performance, it is not possible to replace halon 1301 with HFC-227ea without making hardware changes to the system. In retrofitting either modular or manifold (or banked, distributed) systems, changes to nozzle design and location may be required in addition to modifications to piping.

Replacement of modular systems with higher-capacity and/or additional cylinders poses no particular technical problems. Space could likely be found for larger or additional cylinders in the machinery spaces requiring protection.

Retrofitting manifold (banked) systems would pose a more difficult challenge. This is related partially to the need to markedly increase the quantity of stored agent in bottles located in what may be a confined manifold area rather than being able to place them more easily throughout the protected space proximate to the nozzles, as in the case of modular installations. Further, there would an attendant need for long runs of larger-diameter piping connecting the manifold with the nozzles in order to meet the requisite 10-second discharge time.

For this reason, a hybrid approach has been developed as a way to meet the special retrofit needs of certain manifold ships. This concept involves replacing halon 1301 with HFC-227ea directly in the existing manifold cylinders and also adding modular units with their own nozzles to make up the additional HFC-227ea required. Questions of nozzle location, nozzle design, and the ability of these two systems to function efficiently, particularly with respect to agent distribution and mixing, are as yet unresolved.

A second alternative approach for manifold ships would involve acceptance of some operational risk. Here, if a particular ship class has limited space to accommodate additional HFC-227ea bottles, eliminating the "second shot" capability in those manifold ships now so equipped would make space available for the additional required HFC-227ea.

Water Mist Systems

Retrofitting water mist systems using the current Navy design concept would entail substantial ship modifications. Two pump rooms, for which space may not be available, and the addition of hundreds of feet of pipe would be required. While this might be technically feasible if sufficient space for pump and motor sets were available, the cost associated with the necessary ship modifications, particularly the addition of high-pressure piping, would make ship retrofit with the current generation of water mist systems not feasible overall.

Thus, the cost-effective approach to ship retrofit would be to install HFC-227ea-based gaseous flooding systems that would make maximum use of existing halon 1301 hardware.

Technical Feasibility and Cost

Based on discussions with naval personnel and NAVSEA engineers, the committee believes that retrofitting with HFC-227ea is technically feasible subject to the constraints cited above. This view is corroborated by results of a recent study of four ship classes by the Navy wherein it was determined that retrofit with HFC-227ea was indeed feasible. The cost and the impact on weight and space are shown in Table 4.4.

While the water mist system shows considerable promise for new ship designs, the belief is that HFC-227ea would be the less costly retrofit approach. Further, the Navy technical community feels it is premature to make a judgment as to the retrofitability of a water mist system given its state of relative immaturity, a position with which the committee agrees.

The committee thought a rough estimate of the cost of converting the Navy' s current fleet from halon 1301 to HFC-227ea would be informative. In reviewing the results of the detailed ship studies (see Table 4.4), it was noted that the total cost for hardware, agent, and installation for each ship, arrived at quite independently, was about $82 per pound of currently installed halon 1301, a minimum estimate. Applying this factor to the amount of halon aboard today's ships yields a figure of about $120 million as the cost of conversion from halon 1301 to HFC-227ea. To this must be added the cost of ships not yet built (principally DDG-51s), necessary engineering support and testing, and an estimation reserve. Thus, for $200 million to $300 million the Navy could equip all its ships with non-halon fire extinguishing systems. This figure would of course decrease over time as ships retire from active service.

Table 4.4 Estimated Impacts and Cost for Backfit of HFC-227ea in Selected Ships

SPACE	125-LB HALON	HFC-227 ea PER SHOT	TOT CYL DELTA	CYL WT DELTA	PIPE WT DELTA	SQ FT DELTA	MATL COST ($)	INSTL COST ($)
SHIP: LHA-1[a]								
MMR1	19	48	29	12,760	3,154	45	135,822	118,320
MMR2	23	58	35	15,400	3,806	55	164,072	142,800
AMR	8	20	12	5,280	1,305	19	56,500	48,960
EDGR1	5	13	8	3,520	870	13	36,947	32,640
EDGR2	3	8	5	2,200	544	8	22,822	20,400
JP5 FWD	2	5	3	1,320	326	5	14,125	12,240
JP5 AFT	1	3	2	880	218	3	8,697	8,160
FPR FWD	4	10	6	2,640	853	9	28,250	24,480
TOTAL				**44,000**	**10,875**	**156**	**$467,235**	**$408,000**
TOTAL WT CHANGE (LB)					**54,875**			
TOTAL EST COST								**$875,235**
SHIP: LHD-1[b]								
MMR1	38	48	58	25,520	3,154	91	168,012	236,640
MMR2	46	58	70	30,800	3,806	109	202,922	285,600
AMR	20	25	30	13,200	1,631	47	87,275	122,400
EDGR1	5	13	8	3,520	870	13	36,947	32,640
EDGR2	3	8	5	2,200	544	8	22,822	20,400
JP5 FWD	4	10	6	2,640	653	9	28,250	24,480
JP5 AFT	3 × 60	4	4	852	435	6	12,212	16,320
LCAC PR	3 × 60	4	4	852	435	6	12,212	16,320
HOLD #10	6	15	9	3,960	979	14	42,375	36,720
AVN	2 × 95	4	4	928	435	6	12,644	16,320
1-103-1-A	1 × 15	1 × 60	1	68	109		2,147	4,000
2-H-0-K	4 × 60	5	5	996	544	8	15,913	20,400
2-13-1-K	2 × 10	1 × 60	1	0	109	2	2,147	4,000
2-13-7-Q	1 × 95	2	2	464	218	3	6,322	8,160
TOTAL				**86,000**	**13,920**	**322**	**$652,201**	**$844,400**
TOTAL WT CHANGE (LB)					**99,920**			
TOTAL EST COST								**$1,496,601**
SHIP: CVN-68[a]								
EDGR1	5	13	8	2,560	870	13	36,947	32,640
EDGR2	4	10	6	1,920	653	9	28,250	24,480
JP5 #2	4	10	6	1,920	653	9	28,250	24,480
JP5 #3	3	8	5	1,600	544	8	22,822	20,400
STRM	2 × 15 LB	1 × 95 LB	1	80	0	−2	2,759	4,080
STRM	2 × 15 LB	1 × 95 LB	1	80	0	−2	2,759	4,080
HAZMAT	1	3	2	640	218	3	8,697	8,160
TOTAL				**8,800**	**2,936**	**39**	**$130,484**	**$118,320**
TOTAL WT CHANGE (LB)					**11,896**			
TOTAL EST COST								**$248,804**

SPACE	125-LB HALON	HFC-227 ea PER SHOT	TOT CYL DELTA	CYL WT DELTA	PIPE WT DELTA	SQ FT DELTA	MATL COST ($)	INSTL COST ($)
SHIP: DDG-51[b]								
ER1	20	25	30	13,200	3,263	47	87,275	122,400
ER2	22	28	34	14,960	3,98	53	98,192	138,720
AMR1	10	13	16	7,040	1,740	25	45,827	65,280
GEN RM	6	15	9	3,960	979	14	42,375	36,720
GTM 1A/B	2 × 95	2 × 125	2	88	0	3	4,379	8,160
GTM 2A/B	2 × 95	2 × 125	2	88	0	3	4,379	8,160
SSGTG1	2 × 95	2 × 125	2	88	0	3	4,379	8,160
SSGTG2	2 × 95	2 × 125	2	88	0	3	4,379	8,160
SSGTG3	2 × 95	2 × 125	2	88	0	3	4,379	8,160
FLSR	1 × 60	2 × 95	2	496	109	2	3,869	8,160
FL ISSUE	2 × 15	1 × 95	1	60	0	−2	2,759	4,080
TACTAS	2 × 125	(Deleted-sprinkling)		−880	0	−3	3,000	15,000
TOTAL				**39,276**	**9,788**	**152**	**$305,192**	**$431,160**
TOTAL WT CHANGE (LB)					**49,064**			
TOTAL EST COST								**$736,352**

[a]Technical feasibility and cost confirmed by detailed engineering study.
[b]Technical feasibility and cost estimate based on preliminary study.
SOURCE: Naval Sea Systems Command.

Aircraft Retrofit

HFC-125 is the gaseous agent selected by the military services as an acceptable replacement for halon 1301 in aircraft. As with HFC-227ea in the ship case, a greater weight of agent will be required to achieve the required extinguishing performance—two and one-half to three times the weight of halon 1301 required.

Since inert gas generator technology has not yet reached operational status and such a system must be specifically tailored for each aircraft design, the gas generator is not considered a retrofit candidate at this time by either the committee or the NAVAIR engineers.

NAVAIR provided Table 4.5, which depicts the retrofit potential of HFC-125 into the Navy's current fleet of aircraft. Note that the Navy considers retrofit technically feasible for all planes with the exception of the P-3, and even it could be accommodated if absolutely essential.

Based on the information given in Table 4.5, discussions with naval personnel, and inspection of typical aircraft installations, the committee concurs that HFC-125 can be satisfactorily retrofitted into current aircraft at some penalty in weight. Using a methodology similar to that employed in determining the investment in current halon 1301 systems, NAVAIR estimates that it would cost about $620 million to convert the current fleet, a figure that includes hardware, agent, engineering design, installation, and testing.

Table 4.5 Naval Aircraft Fire Extinguishing Systems—HFC-125 Retrofit Potential

Aircraft Type	Number of Aircraft[a]	Application[b]		Weight (lbs)[c]				Remarks
		Engine	Other	Agent	Bottle	Plumbing	System	
Fixed Wing								
P-3[d]	351	N	Y-APU	50	73	13	135	
F/A-18	826	Y			36	9	62	
A-6E	85	N	N-FUEL					Retires 1997
F-14	422	Y	Y-OWAC	47	76	14	137	
E-2C	123	Y		39	51	4	94	
EA-6B	56	Y		90	90	4	184	
C-2A	38	Y	Y-APU	47	74	4	124	
S-3	134		Y-APU	3	27	0.9	31	
E-6A[d]	16	N	Y-APU	40	87	3	129	
C-130	109	Y		162	138	28	328	
T-44A	45	Y		15	45	6	66	
F/A-18E/F	7	Y		17	36	9	62	EMD only
C-12	80	Y		15	45	6	66	
C-20	7	Y	Y-APU	46	79	3	123	
T-39	14	Y		39	51	6	96	
Rotary Wing								
H-53E	199	Y	Y-APU	54	149	6	209	
SH-60	369	Y		15	60	3	78	
H-53A/D	128	Y	Y-APU	41	110	5	155	
CH-46	253	Y		18	57	5	80	
SH-2	89	Y		15	60	3	78	
H-3	100	Y		15	60	3	78	
AH- 1	213	Y		12	63	3	78	Current upgrade
UH- 1	150	Y		12	63	3	78	Current upgrade
V-22	2	Y	Y-MW	14	61	3	78	FSD only

NOTE: Estimates of platform retrofit weight impact assume a 300% increase in agent and bottle weights and exclude portable fire extinguishers.

[a] The number of aircraft reflect rough estimates, not exact figures.

[b] These columns represent a very rough technical estimate regarding whether a non-ODS system could be volumetrically accommodated. The baseline is a system 300% larger than a halon 1301 system. It does not consider cost or weight factors. There is no HFC-125 retrofit potential in the dry bay application.

[c] Dashes indicate that no retrofit assessment data were available for this analysis.

[d] P-3 and E-6A agent weights include halon 1301 for engine application and HFC-125 for auxiliary power unit application.

SUMMARY, FINDINGS, AND RECOMMENDATION

Halons 1301 and 1211 have served the Navy well as fire extinguishing agents for ships and aircraft since their operational introduction in 1978. Some $1 billion are invested in halon-based fire extinguishing systems hardware, engineering support, testing, installation, and the agent itself, both in platforms or in reserve.

Effective alternative chemical agents have been identified by the Navy and are being incorporated into the design of new ships and aircraft. There is a weight and volume penalty associated with these agents relative to halons, but the impact can be minimized if use of these agents is incorporated into the initial platform design. In addition to these chemical replacement agents, there are promising alternative fire extinguishing systems. The Navy is currently studying and testing water mist and inert gas generator systems and is incorporating these systems into its new-design platforms.

It is technically feasible to retrofit, into existing platforms, non-halon fire extinguishing systems equipped with the replacement chemical agents selected by the Navy for its new-design ships and aircraft. Such a program would cost about $1 billion if executed in the near term, with the amount diminishing over time as ships and aircraft retire from service.

The Navy has sufficient halon 1301 agent in hand to support halon-equipped ships and aircraft until they go out of service. However, inventory projections point to a marginal reserve in the 2030 time period. To hedge against uncertainty, miscalculation, or unanticipated high future usage, the Navy could consider increasing its safety margin by buying recycled halon 1301 in the near term while prices are at a reasonable level. Alternatively, the Navy could consider installing non-halon fire suppression systems in selected new-construction vessels, such as the DDG-51, thereby increasing the halon 1301 reserve by some 400,000 pounds.

FINDING: Effective alternative chemical agents have been identified by the Navy and are currently being incorporated into the design of new ships and aircraft. There is a weight and volume penalty associated with these agents relative to halons, but the impact can be minimized if use of these agents is incorporated into the initial platform design. Further, retrofit of these agents into existing naval platforms is technically feasible in most cases.

FINDING: In addition to the chemical replacement agents, promising alternative fire extinguishing systems such as water mist systems and inert gas generators are under consideration by the Navy for some applications. These systems are being incorporated into new-design naval platforms.

Options

The committee sees several options available to the Navy for meeting its requirements for ongoing, environmentally acceptable effective fire suppression in its ships and aircraft:

1. *Continue on present course.* Continue to implement selected alternative fire protection approaches in new-design platforms. This option is based on the assumption that the current supply of halon 1301 set aside for Navy use will be sufficient for the remaining life of existing ships and aircraft. To hedge against a potential shortfall in the halon 1301 inventory, the Navy could consider buying additional recycled halon to augment the Navy bank and/or adopt alternative agents and technologies in current-design ships not yet constructed, such as the DDG-51. Further, the Navy should maintain, at the present level, its scientific and engineering research effort devoted to developing alternative fire suppression agents and technologies.

2. *Plan for retrofit.* Draft retrofit engineering plans for contingency use to meet the possible need for retrofit of existing ships and aircraft. This option would involve detailed study of retrofit potential and preparation of engineering plans for installing alternative halocarbon agents, water mist systems, and inert gas generators. Given this preparation, the Navy would be in a position to respond quickly if the use of halon were restricted.
3. *Seek the ultimate fire suppression agent.* Fund a major research program directed toward finding a drop-in replacement for halon.

Recommendation

The committee supports continuation of the present course and does not recommend that the Navy underwrite a major new program to seek the ultimate halon 1301 replacement agent.

Appendix A

Halon Use by the Navy

WHY HALON? A HISTORICAL OVERVIEW

Extinguishing fires at sea has always been a matter of priority throughout maritime history, particularly for navies, which are faced with the dual challenge of putting out fires caused by both accidental ignition and enemy action. And crews of ships and aircraft share a common threat from fires—the potential for loss of life. Understandably then, fire prevention and fire fighting readiness are major concerns of captains and pilots as well as senior commanders.

The Navy follows the general practice of categorizing fires by type of fuel—Class A for paper, wood, and general combustibles, B for flammable liquids, C for electrical, and D for metals. The frequency of occurrence and seriousness of a class of fire and the impact on occupants vary considerably between ships and aircraft. A Class B fire, for example, ranks at the top of the critical list and can lead very quickly to loss of an aircraft if unchecked; on a ship the possibility of immediate loss of the vessel is less, but loss of life can be considerable if such a fire is not brought under control in short order.

In recent years the need to prevent loss of aircraft damaged by enemy guns and missiles has been a driver in efforts to develop and install fire extinguishing systems in aircraft. In ships, on the other hand, the requirement to cope with accidental fires, particularly in machinery spaces, has led the way to developing more effective fire extinguishing systems.

Over the years the Navy has sought ways to improve fire fighting capabilities as ships have been required to handle increasing quantities of munitions and volatile aircraft fuels, and have been equipped with propulsion systems requiring high-pressure, easily atomized fuels. Continuing this tradition of evolutionary improvement to meet changing needs, halon was introduced into the Navy as a principal fire extinguishing agent in recognition of its extraordinary fire extinguishing capabilities.

Ships

Fires aboard ship in World War II were fought by damage control teams applying water and protein foam. Hand-held CO_2 extinguishers were widely used, and overhead water sprinklers were also employed in confined spaces and aircraft carrier hangar decks. Steam smothering systems were available in some ships to handle engine room bilge fires. The emphasis in fire fighting was to attack fires directly with men manning hoses dispensing solid water streams or fog, and this remained the accepted approach during the immediate postwar years.

In the late 1960s, a series of aircraft carrier fires incident to Vietnam War operations triggered a search for more effective ways to fight massive flight deck fires. Fighting fires aboard carriers, while always challenging, had become more so in the age of jet aircraft, which carried ten times as much fuel and ten times the weight of explosives as had predecessor aircraft in World War II. As a result, aqueous film-forming foam (AFFF) was introduced as an effective replacement for protein foam, and it became the primary agent in fighting flight deck fires. Flush deck nozzles were installed to dispense AFFF on demand to deal with pooled area fuel fires. Airfield-style fire trucks were put aboard carriers, each carrying AFFF and dry powder (PKP—potassium bicarbonate powder). Additionally, small twin-agent flight deck tractors were equipped with small amounts of AFFF and PKP for quick-reaction fire suppression. Twin-agent hoses also began to appear in ship machinery spaces mounted on reels for ready access by engine room personnel.

Despite these advances, a flammable fuel fire with vertical dimensions remained a fire fighting challenge, as did the pressure-fed engine room fuel spray fire. And in the late 1960s, machinery space fires on board several ships induced a review of both equipment and fire fighting tactics. A variety of technical approaches were considered that would make possible quick evacuation of a machinery space, followed by remote actuation of the fire extinguishing system. After a rigorous selection process and substantial testing in the early 1970s, halon 1301 was chosen as the optimum total flooding agent for "abandon-the-machinery-space" fires.

The first halon systems were installed in aircraft carriers and mine craft in 1978. A policy was established calling for installation of halon in new-construction ships, such as the FFG-7 class frigates, and selective retrofitting into older vessels began on an age-selective basis. The original twin-agent AFFF/PKP reels were retained in engine rooms for small-fire application, but with the PKP side deactivated. Engine rooms also were equipped with AFFF bilge flooding systems. And although the principal reason for acquiring halon 1301 systems was to fight machinery space fires, halon 1301's attraction as a very effective, non-toxic agent resulted in its being substituted for CO_2 in other spaces where flammable liquids were stored.

This introduction of halon to the Navy followed its earlier acceptance for total space flooding applications in the civil community. Thus, because of a confluence of events—availability and civil acceptance of halon 1301, an urgent Navy need for a better agent, and top-level support—halon became the agent of choice for coping with fuel spray fires in confined spaces. Use of halon enabled the Navy to adopt a casualty-reducing tactic of (1) taking the man out of the loop initially by abandoning the fire scene, (2) remotely actuating the halon 1301 flooding system, and (3) reentering the space when the fire was extinguished to deal with any minor residual flare-ups.

Halon 1211 has only limited application aboard ship. It replaced PKP in fire trucks aboard aviation ships in the late 1980s for fighting three-dimensional fires. The agent is also used in mine craft (MSCs) and air cushion landing craft (LCAC), and there are a few hand-held bottles to be found in certain other ship classes.

Aircraft

Fires in aircraft have been a major concern since the inception of powered flight in the early 1900s. The very nature of aircraft—being airborne, carrying large amounts of flammable liquids, containing potential ignition sources—makes them inherently vulnerable to loss if fire should break out. Hence, fire prevention is a major consideration in aircraft design, as are fire extinguishing systems tailored for the specific plane and its anticipated operating environment. Since most fires start in inaccessible areas, particularly in military tactical aircraft, extinguishing them must depend on automatic or remote activation of extinguishing systems. And as mentioned above, combat aircraft have the additional challenge of coping with damage that may be inflicted by enemy antiaircraft artillery and missiles.

Early combat loss experience in World War II highlighted the vulnerability of tactical aircraft to loss by fire and explosion. Self-sealing fuel tanks were installed to reduce the probability of leakage if hit, with the resultant fumes causing explosions in void (dry bay) areas. Additionally, attention was paid to placement of fuel lines and shielding components. CO_2 fire extinguishing systems were installed in the nacelles of multiengine aircraft, as they were in civil airliners of the time.

The introduction of jet aircraft into the Navy in the 1950s was accompanied by a change in strategic emphasis toward nuclear warfare. Attack aircraft were designed to fly long ranges, while designers tried to exact maximum speed and altitude performance from fighters. In the quest for performance, the vulnerability of planes to combat damage, including fire and explosion, was accorded low priority during aircraft design. Even in the case of rotor craft that fly slowly at low altitude, little attention was paid to measures that might reduce vulnerability to loss if the helicopter was struck by enemy projectiles or small missiles.

During the Vietnam War the United States suffered combat losses totaling 5000 aircraft—2500 fixed-wing jets and 2500 helicopters. As losses mounted during the course of the conflict, studies were initiated to see what might be done to lower loss rates, an effort that continued after the war. The analysis revealed that fuel fires and explosions accounted for 50% of the losses and that half of these were attributable to fuel explosions in voids or so-called dry bays. As a result, the military services joined in an effort to improve the survivability of jet aircraft and helicopters. This led to development and experimentation with a variety of approaches that addressed fire extinguishing in several areas, including engine nacelles/bays, dry bays, fuel tanks, occupied spaces, and ground and ship flight decks. Technologies considered were (1) solid foams, powders, and inert gas generators for dry bays; (2) solid foams and inert gases for fuel tank ullage areas; (3) halon 1301 for engine nacelles and bays; (4) portable halon bottles, principally 1301, for occupied areas; and (5) AFFF and halon 1211 for crash fire fighting and small fires incident to engine start.

The adoption of halons 1301 and 1211 was the culmination of fleeting military involvement with halons over the years. In the 1920s non-fluorinated halon agents were tried experimentally in engine nacelle extinguishers, but their use was abandoned by the U.S. military in favor of the non-corrosive CO_2. Despite their relatively high inhalation toxicity, systems using halon 104, 1001, and 1011 were developed during World War II and employed by the British and Germans in military aircraft. The use of these agents expanded into the civil sector after the war. In the United States, however, it was only after development of fluorinated halons (1301, 1211) that CO_2 was replaced in Air Force and Navy aircraft by these new highly effective, less toxic, and non-corrosive agents. And since they had already gained some acceptance in U.S. civil aviation as well as in various civil ground applications, the military quickly adopted them to meet the variety of needs cited above.

SHIP FIRE EXTINGUISHING SYSTEMS

Fire at sea has always posed a special danger. In warships, the fire hazard is exacerbated by the threat of explosive weapon warheads and propellants. Throughout its history the Navy has dealt with fire protection challenges by exacting the most from existing fire fighting systems through organization and training as well as by exploiting new technologies. The exploitation of dry chemical powders and aqueous film-forming foam as well as the introduction of specialized naval fire fighting systems are examples of the constant improvement sought by the Navy in the safety and survivability of its vessels, aircraft, and crews. Employing halons for machinery space and aviation fire extinguishing applications is an example of adopting new technology to improve fire protection.

Machinery Space Fires

The principal fire threats in machinery spaces are the combustible liquid pool and pressurized spray fire. The most hazardous type of incident, and one that absolutely requires a gas-phase fire suppressant, is the three-dimensional spray or cascading fire. These fires arise from fuel or lubricant pipe or fitting leaks, human error, or mechanical damage. Leaks can vary in scale from less than 1 to greater than 50 gallons per minute. Pressurized spray fires generally occur in fuel, lubricating oil, or hydraulic fluid system piping. Pressures range from 10 to 1000 psi. Non-pressurized cascading fuel fires often involve sounding tubes, gravity storage tanks, and fuel piping that transit the space servicing other areas such as aviation fuel systems. In general, a spray or cascading fuel fire will also produce a pool fire.

A release of fuel or lubricating oil can be quickly ignited by hot surfaces (steam pipes or boiler fronts), electrical arcing or shorts, welding operations, and mechanical sources (friction, sparking, and so on, related to equipment failure). The intensity of these fires can easily approach 50-MW power equivalent. The fire growth time scale is on the order of several seconds, so that very large fires, high temperatures, and fatal concentrations of carbon monoxide (CO) can occur in 30 seconds or less. Since there is insufficient oxygen to maintain a large fire, the power level will decrease with time, and higher CO production will occur.

The size and growth rate of these three-dimensional fires preclude safe reliance on manual firefighting in closed spaces. Clearly, manual fire fighting against a large machinery space fire is not the approach of choice because of the rapidity with which the space becomes untenable due to heat, smoke, and toxic combustion by-products. Indeed, these hazardous conditions are the very reason halon 1301 was introduced in the Navy some 20 years ago.

Reignition is also a key consideration in fighting machinery space fires. Three sources of reignition in ship machinery spaces—hot surfaces, electrical sources, and smoldering solid combustibles—form the basis for the required agent hold times and reentry procedures. Each source of reignition is described briefly below.

- *Hot surfaces*—These result from normal power generation in a ship and include steam piping where lagging is breached, diesel engine exhaust manifolds, gas turbine casing and exhaust stacks, and boiler fronts. Some hot surfaces, particularly steam piping, can remain above the fuel autoignition temperature for hours. Unless there is an exceptionally long (> 5 to 10 minutes) preburn time associated with a fire, metal surfaces heated to above the fuel autoignition temperature will cool relatively quickly after the fire has been put out. Accordingly, after a typical 20- to 30-minute agent hold/cool down period, fire-heated surfaces will usually not be a hazard.
- *Electrical sources*—These arise from fire damage to electrical equipment and cabling. Shorts between cables or to ground may result in arcing or resistance heating. These reignition sources, if in proximity to a fuel surface or fuel vapor, are energetic enough to cause reignition and, in extreme cases, deflagration of a fuel vapor cloud. These reignition sources will remain a threat until all the power to the space is secured. While it is a relatively straightforward exercise to secure power serving a space, it is difficult in some ships to secure power in all cables that transit the space.
- *Smoldering fires*—Some transient combustibles, insulation, and lagging materials are susceptible to smoldering combustion. Typical design concentrations of halon 1301 (or replacements) are not sufficient to extinguish the condensed-phase, oxidation process, particularly in cellulosic materials. Thus, once the agent concentration has decayed sufficiently, a smoldering fire may reignite.

Extinguishing Systems

Three systems are employed to control or extinguish fires in machinery spaces. The total flooding halon 1301 system is the key element of a three-element overall system of fire protection. It is capable of extinguishing any flammable or combustible liquid fire, pool or spray, as well as solid combustibles ignited as a result of the liquid pool or spray fire. Halon 1301 is used when a fire is too large to suppress manually, which is usually the case with pressurized spray fires.

Additional fixed protection in machinery spaces is offered by the AFFF bilge foam system. It is designed to extinguish pool fires caused by fuel or lubricating oil leaks and to prevent reignition. The system may be used in conjunction with halon 1301 or independently to extinguish and secure fuel spills in the bilges.

A third means of suppression is to fight a fire manually with portable extinguishers and hose streams. Hose streams include AFFF hand lines as well as regular water-only hoses available throughout the ship. Manual fire fighting may be employed in machinery spaces (1) when a fire is small or localized and can be readily extinguished by watch standers without protective equipment and (2) when a space is reentered after a halon system has been activated in order to extinguish any residual fire or, less likely, if the application of halon has not been completely effective.

Halons

Halon 1301 and halon 1211 are employed on board ships to meet unique fire fighting challenges—where their fire extinguishing effectiveness, low toxicity, minimum space and weight requirements, and lack of agent-induced damage are required characteristics.

Halon 1301 is used in machinery spaces and flammable liquid storage and issue rooms because of its effectiveness against those fires requiring total gas flooding, low toxicity, lack of agent-induced damage, and a relatively low system space and weight impact. This agent is suitable for flammable gas, liquid, and typical solid combustible fires. It extinguishes fires in enclosed spaces by employing the principle of gas-phase catalytic interruption of combustion reactions (see Table 2.1), when used in concentrations of 5 to 6% by volume. A properly designed system will distribute a uniform concentration throughout the space, thereby providing extinguishment at all locations. Once a uniform mixture of halon in air is generated, the extinguishing concentration must be maintained for a period of time to preclude possible reignition.

Halon 1211 is used for so-called streaming or local applications where it is desirable to direct a stream of liquid agent to a localized fire. In Navy applications, halon 1211 is always manually applied and directed at a specific fire location. On board aviation ships, it is used to extinguish small fires in engines that result from the pooling of fuel when an aircraft engine fails to start. It is also employed in fighting large three-dimensional cascading flight deck fires. Halon 1211 is applied from portable extinguishers, wheeled bottle carts, and crash vehicle hose lines. Finally, halon 1211 is employed in LCAC engine compartments for fire suppression.

Aqueous Film-Forming Foam

Aqueous film-forming foam is a mixture that is 6% AFFF concentrate (primarily fluorosurfactants and solvents) and 94% water, either seawater or fresh water. It is used to extinguish pool fires and to prevent the fuel vaporization and subsequent reignition. All foams, including AFFF, are only effective in suppressing two-dimensional flat pools of fuel and are generally ineffective on spray fires or cascading fuel fires.

The primary shipboard uses of AFFF are in machinery spaces, aircraft hangars, fueled vehicle stowage areas, and on flight decks. In machinery spaces, an AFFF spraying system is located in the bilge to extinguish fires and secure fuel spills to prevent ignition. AFFF hose lines are also provided for manual application. Flight deck applications include an AFFF spraying system with flush mounted nozzles located in the flight deck. AFFF hoses are also provided on the flight deck and on crash/fire vehicles.

While AFFF is a primary flight deck extinguishing agent, halon 1211 is used to attack cascading or spray fires which may occur as a result of a crash or catastrophic failure, and for small engine or "wet start" fires. In most cases, AFFF is used to extinguish and secure the pool fire resulting from a crash or large wet-start, and halon 1211 is used to extinguish localized three-dimensional fuel fires.

Other Fire Fighting Measures

In addition to the special-hazard extinguishing systems and agents described above, the Navy uses traditional fire protection systems as appropriate. On new-construction ships, automatic sprinkler systems are installed in storage and berthing spaces as well as in areas surrounding vital electronic equipment, and water deluge systems are provided for ordnance magazine cooling.

With respect to manual fire fighting equipment, seawater fire mains run throughout a ship, with hose stations localized so as to provide coverage in all areas. Portable extinguishers, filled with CO_2, AFFF, or PKP are distributed throughout the ship for first-response fire fighting.

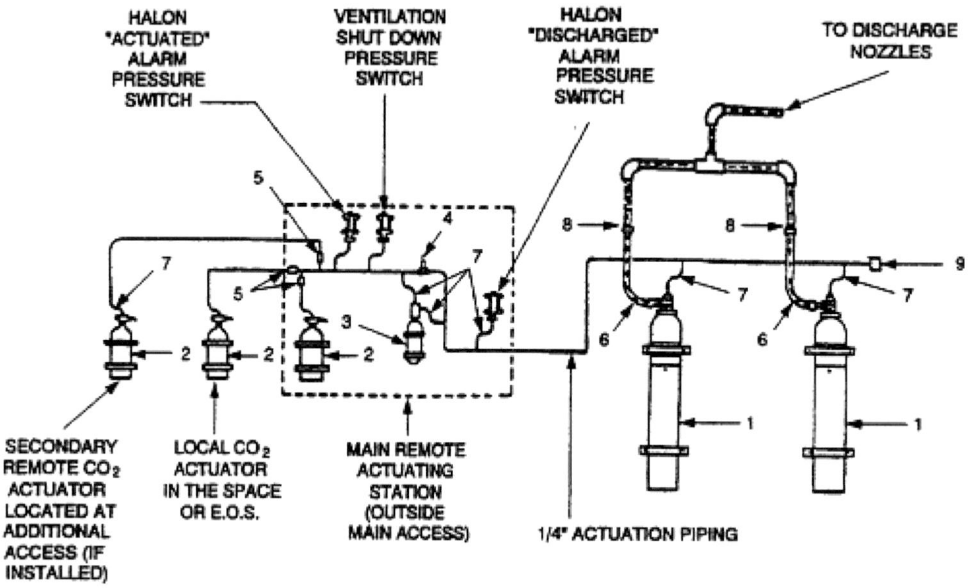

ITEM NO	COMPONENT
1	HALON CYLINDER
2	MANUAL CO ACTUATOR
3	TIME DELAY DEVICE
4	TIME DELAY BYPASS VALVE
5	1/4 " CHECK VALVE
6	1-1/2 " FLEXIBLE HOSE
7	1/4 " FLEXIBLE HOSE
8	1-1/2 " CHECK VALVE
9	VENT FITTING

NOTES:

1. 1/4" ADAPTER FITTINGS ARE NOT SHOWN. LOCATION IS BETWEEN 1/4" PIPING AND 1/4" FLEXIBLE HOSE (7).
2. QUANTITY AND SIZE OF HALON CYLINDERS VARIES FOR EACH INSTALLATION.
3. 1-1/2" CHECK VALVE MAY NOT BE INSTALLED WHEN THERE IS ONLY ONE CYLINDER.

———— = CO_2 FLOW

.... = HALON FLOW

Figure A.1
Typical distributed (banked) halon system.

How a Shipboard Halon 1301 System Works

Halon 1301 systems installed in shipboard machinery spaces are designed to discharge sufficient halon into a space to develop a uniform 5% (by volume) concentration of halon 1301 in air throughout the space. Two types of systems are employed—manifold and modular. Manifold systems, also called "banked" or "distributed," consist of a bank of cylinders connected through a manifold or distribution piping (Figure A. 1). Piping from the bank runs throughout the space to nozzles located on the overhead and beneath the intermediate deck levels. Manifold systems were, and still are being, installed in new ships built after 1980. Modular systems feature individual cylinders of halon distributed throughout the protected space, connected via short lengths of pipe to one or two nozzles nearby (Figure A.2). These systems were used principally to retrofit existing ships with a halon 1301 system at the time the decision was taken to install halon fire extinguishing systems in naval vessels.

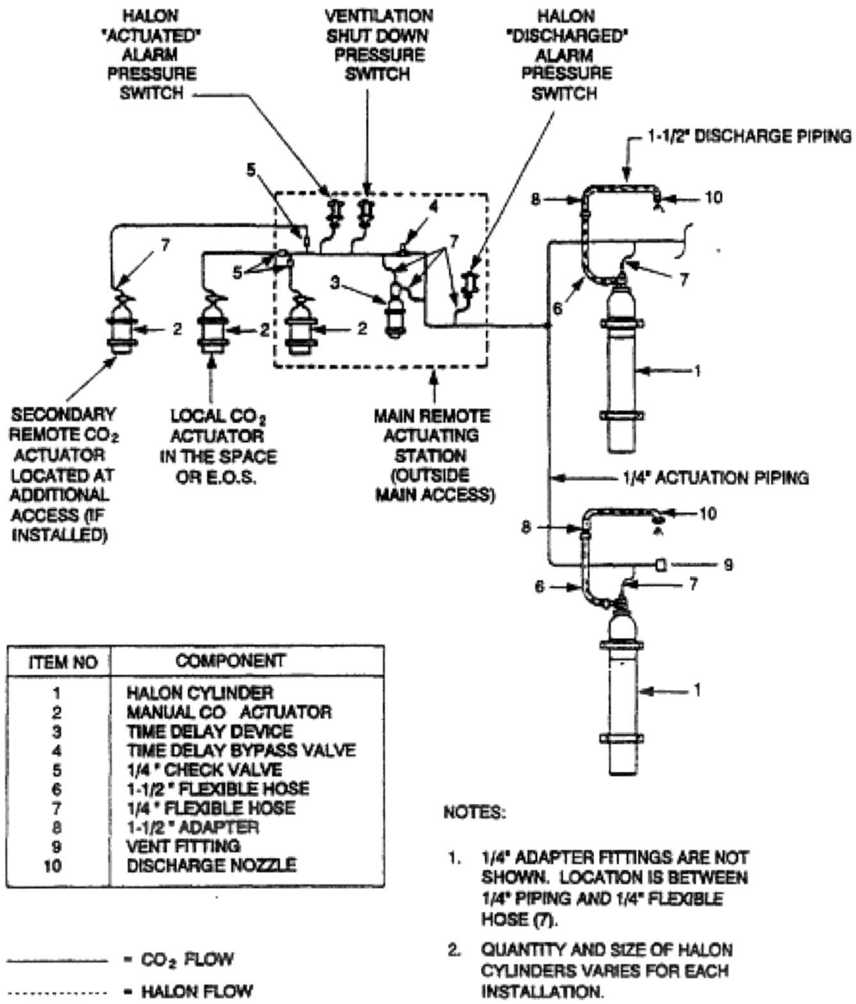

Figure A.2
Typical modular halon system.

The halon 1301 agent is in liquid form in cylinders, pressurized by nitrogen at 600 psi. The nitrogen is required to provide adequate cylinder pressure to discharge the halon in 10 seconds and to provide sufficient nozzle pressure to ensure proper distribution of the agent in the compartment.

All machinery space halon systems are manually actuated using pneumatic actuation lines. Actuation stations are located inside the space, at least one of which is near the normal exit. Additionally, another station is outside the space, often on the main deck above proximate to the path leading to and from the space. Actuation of a station trisects release of CO_2 from a small cylinder. This release in turn actuates a pressure switch that shuts down the ventilation fans to the space and closes dampers if installed. The CO_2 flow also passes through a time-delay device. Warning horns are sounded during the 30- to 60-second time delay, signaling that the affected space is to be evacuated prior to halon discharge. After approximately 30 to 60 seconds, CO_2 is permitted to flow to the halon cylinder actuation valves, thus initiating halon release.

For manifold systems, a single, small CO_2 cylinder opens valves for all cylinders coupled to the manifold. Modular systems, on the other hand, require pneumatic actuation piping to each cylinder for halon release. In either case, halon is discharged from the cylinders within approximately 10 seconds after the valves are opened.

Once the halon has been released, fire fighting doctrine calls for monitoring the space for 20 to 30 minutes. In certain manifold system ships, a second halon discharge is available if needed—the so-called "two-shot" system. After 30 minutes, a fire party reenters the space to extinguish residual fires and provide cooling with AFFF hoses if necessary.

Shipboard Spaces Where Halon 1301 Systems Are Installed

Tables A.1 to A.5 list the location of halon 1301 extinguishing systems in five representative ship classes—two amphibious aviation ships (LHA, LHD), two surface combatants (FFG-7, DDG-51), and an aircraft carrier (CVN-73). It can be seen that halon is installed in those spaces where flammable liquids are used for power generation, where such liquids are stored, and where aviation fuel is handled.

In general, older ships have one-shot modular systems since these were easier and less costly to retrofit than the two-shot manifold (banked) installations characteristic of vessels that had not yet begun construction at the time of the decision to install halon 1301 fleetwide. Notable is that the large CVN has less halon installed than a small, 3650-ton FFG-7 frigate. This is attributable to the nuclear-powered CVN being a one-shot ship, but more importantly, to the lack of a requirement for any halon in main propulsion spaces. It also highlights the point that main machinery spaces, emergency diesel generator rooms, and aviation fuel pump rooms account for a large proportion of the halon installed in applicable ship classes.

Table A.1 LHA-4 USS Nassau Class Halon 1301 Systems

Space	Cylinders	Size (lb)	Halon Quantity per System (lb)
Main Machinery Room #1	19	125	2,375
Main Machinery Room #2	23	125	2,875
Auxiliary Machinery Room	8	125	1,000
Emergency Diesel Generator Room # 1	5	125	625
Emergency Diesel Generator Room #2	3	125	375
JP-5 Pump Room	2	125	250
JP-5 Pump Room	1	125	125
Fuel Pump Room	4	125	500
Quantity of halon installed on ship			8,125
On-board spares			3,250
Total halon on board			11,375

NOTE: All systems are "single shot" and "modular." This arrangement (single shot, modular) is typical of ships that received halon via backfit.

Table A.2 LHD-3 USS Kearsarge Class Halon 1301 Systems

Space	Cylinders	Size (lb)	Halon Quantity per System (lb)
* Main Machinery Room #1	38	125	4,750
* Main Machinery Room #2	46	125	5,750
* Auxiliary Machinery Room	20	125	2,500
Emergency Diesel Generator Room #1	5	125	625
Emergency Diesel Generator Room #2	3	125	375
JP-5 Pump Room #1	4	125	500
JP-5 Pump Room #2	3	60	180
LCAC Pump Room	3	60	180
Paint Mix & Issue Room	1	95	95
Cargo Flammable Liquid Room	6	125	750
Aviation Flammable Storeroom	2	95	190
Supply Department Flammable Storeroom	4	60	240
Ship Store Flammable Liquid Storeroom	2	10	20
Aviation Flammable Liquid Storeroom	1	15	15
Quantity of halon installed on ship			16,170
On-board spares			2,050
Total halon on board			18,220

NOTE: Entries with an asterisk (*) have "two-shot" systems. All systems are "banked."

Table A.3 FFG-7 USS Perry Class Halon 1301 Systems

Space	Cylinders	Size (lb)	Halon Quantity per System (lb)
* Engine Room	14	95	1,330
* Auxiliary Machine Room #1	4	95	380
* Auxiliary Machine Room #2	18	60	1,080
* Auxiliary Machine Room #3	8	60	670
* Emergency Diesel Generator Room #1	2	95	190
* Emergency Diesel Generator Room #2	2	95	190
* Emergency Diesel Generator Room #3	2	95	190
* Emergency Diesel Generator Room #4	2	95	190
* Gas Turbine Module 1A	2	60	120
* Gas Turbine Module 1B	2	60	120
Flammable Liquid Storeroom	3	10	30
Flammable Gas Cylinder Storeroom	1	95	95
Paint Mix & Issue Room	2	10	20
* TACTAS Room	2	95	190
* RAST Machinery Room	2	60	120
Quantity of halon installed on ship			4,915
On-board spares			495
Total halon on board			5,410

NOTE: Entries with an asterisk (*) have "two-shot" systems. All systems are "banked."

Table A.4 DDG-51 USS Arleigh Burke Class Halon 1301 Systems

Space	Cylinders	Size (lb)	Halon Quantity per System (lb)
Auxiliary Machinery Room	10	125	1,250
Engine Room #1	20	125	2,500
Engine Room #2	22	125	2,750
Generator Room	6	95	570
Gas Turbine Module - 1A/B	2	95	190
Gas Turbine Module - 2A/B	2	95	190
Ship Service Gas Turbine Generator #1	2	95	190
Ship Service Gas Turbine Generator #2	2	95	190
Ship Service Gas Turbine Generator #3	2	95	190
Flammable Liquid Storeroom	1	60	60
Flammable Liquid Issue Room	2	15	30
TACTAS Room	2	125	250
Quantity of halon installed on ship			8,360
On-board spares			635
Total halon on board			8,995

Table A.5 CVN-73 USS George Washington Class Halon 1301 Systems

Space	Cylinders	Size (lb)	Halon Quantity per System (lb)
Emergency Diesel Generator Room #1	5	125	625
Emergency Diesel Generator Room #2	4	125	500
Pump Room #2	4	125	500
Pump Room #3	3	125	375
Flammable Liquid Storeroom #1	1	125	125
Flammable Liquid Storeroom #2	3	95	285
Flammable Liquid Storeroom #3	2	95	190
Aviation Flammable Storeroom	2	125	250
Paint Mix & Issue Room #1	1	60	60
Paint Mix & Issue Room #2	2	15	30
Aviation Paint Mix & Issue #1	2	15	30
Aviation Paint Mix & Issue #2	1	15	15
Bomb Hoist Room #1	2	15	30
Bomb Hoist Room #2	2	15	30
Quantity of halon installed on ship			3,045
On-board spares			1,295
Total halon on board			4,340

AIRCRAFT FIRE EXTINGUISHING SYSTEMS

Halon 1301 fire extinguishing systems in aircraft are employed in ways related to the size and mission of the aircraft and the number of engines. Application areas include engine nacelles or engine bays of twin-engine tactical aircraft; dry bays—those void spaces adjacent to, or beneath, fuel tanks and fuel lines; portable, hand-held extinguishers in cockpits and cabin spaces where aircrew personnel are located; and miscellaneous uses such as inerting a fuel tank's ullage (space above fuel in a less-than-full tank) to prevent incendiary bullet-initiated explosions, and to extinguish fires in auxiliary power units.

Fires Aboard Aircraft

Aircraft applications involve suppression of four distinct types of fires. These include hydrocarbon/air diffusion flames characteristic of engine nacelle and bay fires; premixed fuel vapor/aerosol/air deflagrations in dry bay applications; explosions of premixed fuel/air mixtures in fuel cells; and solid combustible diffusion flames involving cable and wire insulation or other combustibles that are typically suppressed using hand-held portable extinguishers.

The most critical fires are those in engine bays and explosion/deflagration events in dry bays. Each of these situations is addressed briefly below.

Halon 1301 is now used extensively to protect engine nacelles/bays. Here, the agent is discharged at a high rate through a series of nozzles to mix with the air stream through the engine and form a transient extinguishing concentration as it passes through the nacelle/bay and extinguishes the in-flight fire. The agent is discharged on command of the pilot after a positive indication of an engine fire, usually from a combination of thermal fire detection activation and/or anomalies in engine operating parameters. Thus, while the agent is discharged in only a few seconds, tens of seconds may elapse between the first indication of trouble and discharge of the suppression agent.

The basic mechanism of extinguishment of engine bay fires is identical to suppressing any hydrocarbon/air diffusion flame, such as a fuel oil fire in a ship machinery space. The primary technical challenge in aircraft is to disperse a sufficiently high concentration of agent into the engine bay such that an extinguishing concentration is maintained within the very high air flow environment of the bay. Generally, this extinguishing concentration is maintained for only a few seconds. Given the high air velocities present in engine bays, the flame strain rate is much higher than what occurs in quiescent, buoyant diffusion flames. As a result, these flames can be extinguished at lower agent concentrations. The reduced partial pressure of oxygen at flight altitude also simplifies the suppression process. A complicating factor in system design is the highly obstructed nature of the engine bay, which impacts nozzle design, flow rate, and agent mixing behavior.

The suppression of explosions (inertion) in dry bay applications requires sensing the presence of an explosion kernel before the flame front has expanded to a damaging size, and then rapidly applying a suppression agent in the vicinity of the ignition to quench the deflagration wave. This sequence must occur within a time scale of tens of milliseconds in order to effectively limit explosion damage, hence the need for automatic system actuation.

The primary scenario for initiation of a dry bay explosion is ordnance or shrapnel penetrating a fuel cell adjacent to the bay and subsequent ignition of the resulting fuel/air mixture. Dry bay explosion suppression systems are designed primarily to ensure the survivability of aircraft in combat. An alternative, albeit much less efficient approach, is to inert bays and voids prior to combat damage. This approach has been tried with some aircraft but has not been pursued in more recent aircraft designs.

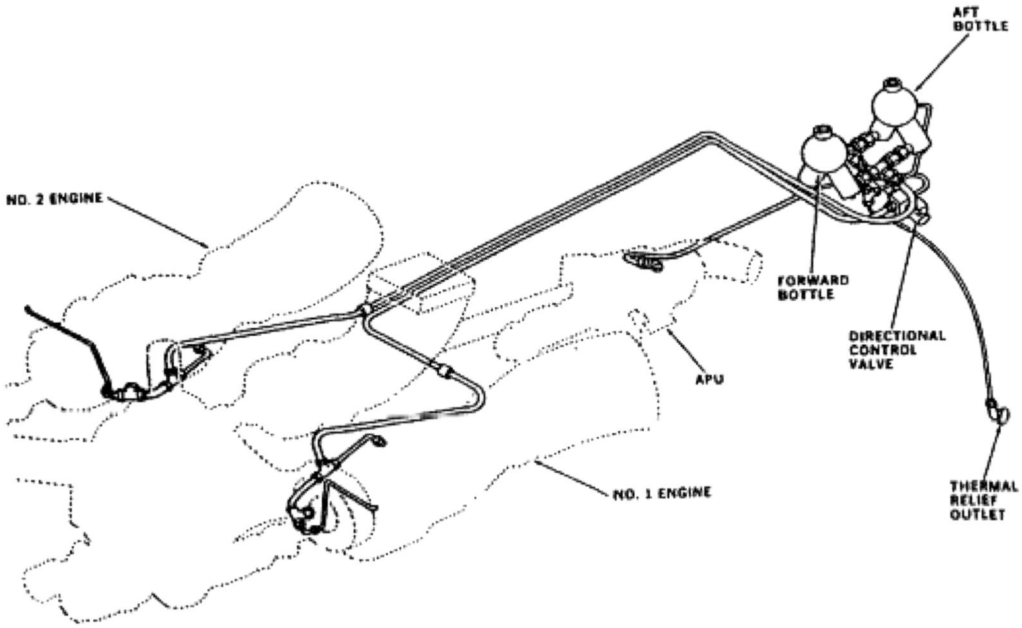

Figure A.3
Typical halon aircraft engine fire extinguishing system.

Typical Halon 1301 Aircraft Engine Fire Extinguishing System

A typical halon 1301 aircraft engine extinguishing system is composed of fire detection sensors linked to cockpit warning lights, halon bottle(s) pressurized by nitrogen at 600 psi, tubing from bottles to strategically placed nozzles, and a pilot-actuated linkage (mechanical, electrical, pneumatic) connecting the cockpit to the halon bottle(s). See Figure A.3.

It should be noted that no single-engine naval aircraft has fire extinguishers in the engine bay. This is based on the premise that, should a fire start because of battle damage or a severe fuel leak and be extinguished, it makes little sense to restart the engine after having once cut off the fuel that was feeding the fire.

Dry bays or void areas alongside or beneath fuel tanks, and through which fuel lines may pass, are susceptible to explosions and fires if combat damage is suffered. Protection measures employed include solid foams, inert gas generating systems, and halon 1301. In such an application, a halon 1301 system would activate automatically in milliseconds upon detection of an explosion kernel by optical flame detectors.

Appendix B

Regulation of Halon and Halon Replacements

When stratospheric ozone is depleted by halons and other ozone-depleting substances, increased amounts of harmful ultraviolet radiation reach Earth's surface. Increases in UV-B radiation are likely to have substantial adverse effects on human health, including increases in the incidence of, and morbidity from, skin cancer, eye diseases, and infectious diseases.[1] Peak global ozone depletion is expected to occur during the next several years, and the stratospheric ozone layer is expected to recover in about 50 years in response to international actions under the Montreal Protocol and its amendments and adjustments.[2] The early phaseout of halon production accounts for at least 15% of the protection provided under the Montreal Protocol.[3]

The phaseout of halon production took effect January 1, 1994, with little disruption because the fire protection community had established global information networks and coordinated halon banks. Halon banks are important because environmentally acceptable alternative extinguishing agents have not been commercialized for some critical fire protection applications (15 to 20% of former uses).[4,5] The success of a production ban on halons is predicated on the free exchange of existing halons, the open use of recycled halons, and a safety valve to allow for production should the banking scheme fall short of expectations (Decision IV/25 of the Montreal Protocol, which allows for continued production for "essential" uses).

In 1985, a small group of countries signed the Vienna Convention on Ozone Layer Protection, the framework for negotiating the Montreal Protocol. In that document, halons are mentioned only briefly in an annex on monitoring of data, because earlier analysis had concluded that halon was rarely released and had predicted that halon use would decline as computer systems became smaller. In 1986, few substitutes had been identified for any of the ozone-depleting fire extinguishing substances, and it was widely believed that halon uses were all essential. It was hoped that chlorofluorocarbon (CFC) restrictions alone would adequately protect the ozone layer.

By late 1986, the U.S. Environmental Protection Agency (EPA) had begun to examine the extent of halon use. The National Fire Protection Association (NFPA) planned to mandate full discharge testing of all new halon 1301 (CF_3Br) systems in order to verify that the controls and hardware functioned properly and that the concentration of halon gas was high enough and remained long enough in an enclosure to extinguish a test fire. EPA was concerned that property owners, insurance companies, and fire authorities might also conclude that older systems should be discharge tested or that all systems should be periodically discharge tested. Such testing alone would have substantially increased the threat to the ozone layer.

Because halons were not part of any regulatory plan and because fire protection involved human life and property, EPA officials met with the chair of the NFPA halon 1301 committee to discuss collaborative efforts to investigate halon use. It was estimated that very little halon was used to actually fight fires but that emissions from testing, training, and accidental discharge were far higher than analysts had thought. A plan was proposed to involve global experts in problem solving and to use market incentives to change the way that engineers and property owners protected against fire risk. It was agreed that EPA and the fire protection community jointly should investigate halon controls, with the goal being to act only by broad consensus.

In early 1987, EPA initiated projects with the U.S. Department of Defense and U.S. Air Force. The Air Force sent a representative to Montreal to help make the case that halon should be included in the protocol. Diplomats reasoned that if the military could reduce its use, so could the civilian sector. Without this endorsement, halon production might not have been included in the 1987 Montreal Protocol.

Further analysis indicated that less than 10% of halon emissions were attributable to fire fighting.[6] EPA, NFPA, and other organizations were now working to educate stakeholders about the importance of eliminating testing, training, and accidental discharges. In Australia, the State of Victoria implemented strong controls on halon use, and plumbers' unions refused to install or service halon systems unless it was deemed essential by a committee of public and private experts. Elsewhere, authorities of jurisdiction were helping to eliminate requirements for discharge testing and training with halon.

In 1989, the United Nations Environment Programme (UNEP) organized the first technology assessment, which included the work of the UNEP Halon Technical Options Committee. This committee of international experts became the catalyst for global efforts.

Slowly fundamental change began to occur. Property owners began to use a broader range of strategies to protect property. Computer manufacturers confirmed that, contrary to advertising claims for halon, most equipment could be protected with water sprinklers. Insurance companies agreed to offer their most favorable rates to insure property with fire protection other than halon. Telecommunications companies reduced the need for halon by using cable materials that would not bum. The military began to design weapons systems that did not depend on halon. Broader fire protection engineering considerations and fire prevention began to take precedence over the basic fire-extinguishing perspective. These efforts stimulated other important paradigm shifts. For example, military aircraft designers reevaluated whether space and weight might be better allocated to threat avoidance or weapons rather than fire protection.

The EPA and the Air Force helped to organize the Halon Alternatives Research Corporation (HARC) to aid in identifying the most promising research opportunities, and they worked to prepare markets to accept alternatives and substitutes as they developed. The Marine Corps, Navy, and Air Force cooperated to develop the first practical halon recycling equipment and were the first organizations in the world to deploy this equipment. The Navy and Marine Corps teamed up with EPA to teach halon recycling to experts from Latin America and the Far East.

Unfortunately, halons are still required for 15 to 20% of the applications they satisfied in 1986. If halons currently contained in existing equipment are never released to the atmosphere, the integrated effective future chlorine loading above the 1980 level is predicted to be 10% less over the next 50 years.[7] See Chapter 3 for further discussion.

Thus, much work remains to complete the phaseout of halon use. Chemical substitutes for halon for the remaining important uses are a part of the ultimate solution.

U.S. REGULATION OF HALONS AND HALON SUBSTITUTES

When the Montreal Protocol was signed in 1987, the EPA's role in stratospheric ozone protection derived from the Clean Air Act of 1977, Part B, section 157(b):

> . . . the Administrator shall propose regulations for the control of any substance, practice, process or activity (or any combination thereof) which in his judgment may reasonably be anticipated to affect the stratosphere, especially ozone in the stratosphere, if such effect in the stratosphere may reasonably be anticipated to endanger public health or welfare.

This language gave EPA broad latitude, but it did not give clear guidance. EPA began to develop control strategies based primarily on measures of ozone depletion potential (ODP). A product whose ODP was lower than that of the CFCs was considered to have an advantage over the halons. Thus, FM-100™ (HBFC-22B1 or CF_2HBr) with an ODP of 0.74[8] was investigated as an effective halon substitute. With the enactment of the Clean Air Act Amendments of 1990 (CAAA), Congress provided guidance to EPA by stipulating that any substance with an ODP of 0.2 or greater would be a class I substance and would be subject to the same production phaseout as the CFCs and halons. This restriction effectively eliminated some potential fire-extinguishing substitutes, such as FM-100™, and mixtures using CFCs.

Title VI of the U.S. Clean Air Act of 1990 enacts the U.S. strategy for complying with the Montreal Protocol for protection of Earth's stratospheric ozone layer.[9] Title VI is administered by the Stratospheric Protection Division within the Office of Air and Radiation. Section 612 of Title VI directs EPA to set up the Significant New Alternatives Policy (SNAP) program, to evaluate any halon substitutes or alternative fire protection technologies to ensure that they reduce the overall risk to human health and the environment and to promote these substitutes to achieve rapid market acceptance. EPA's goal is to ensure that industry and consumers have ample choices for the diversity of applications in which CFCs and halons are currently used.

EPA adopted a risk-balancing approach on health and safety issues by looking at likely exposure pathways in use of a substitute agent in each sector. The risk to individuals from exposure to halon substitutes is generally from discharges that occur infrequently. Chronic effects from exposure to halon substitutes are not usually a concern, because when used, these substances are discharged in high concentrations over short periods of time and thus potentially present an acute hazard. Risk from exposure to halon substitutes is accordingly best assessed by analysis of acute toxic effects associated with exposure to these compounds, such as developmental toxicity and cardiotoxicity. In most instances, cardiotoxicity occurs at lower levels than does fetotoxicity, and therefore, unless otherwise warranted by the developmental data, EPA bases the estimates for emergency exposure limits during halon use on the no observable adverse effect level (NOAEL) and lowest observable adverse effect level (LOAEL) reported for epinephrine-sensitized cardiotoxicity in dogs (and in a few instances monkeys). Human heart arrhythmias and sudden death resulting from overexposure to CFCs, halons, and other halogenated and non-halogenated hydrocarbons have been documented in work-place settings and in volatile substance abuse (e.g., glue sniffing).[10]

To assess the safety of a fire extinguishing agent for use in a total flooding system, EPA analysts examine the actual design concentration as NFPA defines it,[11] i.e., the cup burner extinction concentration plus 20%, or in some cases the actual large-scale testing design concentration, and compare this value to levels at which cardiotoxic effects have been observed.

The situation differs for streaming agents (i.e., chemicals applied to localized fires, usually by being propelled from an extinguisher) because such use is a localized application, and air exchange further dilutes the concentration of the agent. EPA requires manufacturers to submit data acquired by personal monitoring for the anticipated usage. The results of these tests show that actual exposure is much lower than what the models predict. Consequently, EPA has listed agents as acceptable, even with a LOAEL as low as 1.0 or 2.0%.[12]

The conditions stipulated under SNAP for use of total flooding agents are patterned after current Occupational Safety and Health Administration (OSHA) requirements for use of halon 1301 (CF_3Br) systems. Because OSHA does not currently specify acceptable levels of exposure to substitute fire extinguishing agents, EPA is laying these values out very specifically and has initiated efforts to work with OSHA as that agency takes steps to amend its regulation of fixed gaseous extinguishing systems (OSHA Regulation 1910.162).

When considering environmental effects of halon substitutes, EPA first looks at ozone depletion potential to determine if a substance could significantly damage the stratospheric ozone layer. Any class I substance (ODP of 0.2 or higher) must be phased out of production in the United States within 7 years of listing. While the Clean Air Act does not explicitly define a class II substance, by implication it is an agent with an ODP of less than 0.2. Currently the chemical with the lowest ODP that EPA has listed as a class II substance is HCFC-123 (CF_3CHCl_2), with an ODP of 0.02.

While EPA considers other environmental impacts besides ozone depletion potential (including aquatic toxicity, air pollution, and so on), global warming potential (GWP) and atmospheric lifetime are the key additional issues in evaluating halon substitutes. Action number 40 of President Clinton's Climate Change Action Plan, released in November 1993, directs EPA to minimize unnecessary emissions of greenhouse gases to help meet the national goal of reducing emissions in the year 2000 to 1990 levels. EPA again has adopted an approach that seeks to balance the risk posed by ODP and GWP and the related atmospheric lifetimes of these agents.

The CAAA directs EPA to "reduce overall risks to human health and the environment."[13] EPA has attempted to characterize emission levels and exposure pathways in each use sector in order to minimize environmental impacts. Thus, EPA first looks for the outliers such as the perfluorinated carbons (PFCs), which have atmospheric lifetimes in excess of 3,000 years and are virtually indestructible.[14] Yet, because PFCs have a favorable toxicity profile, EPA recognizes that they can play a role in fire protection applications where other agents are not suitable for either technical or safety reasons. Thus, EPA has listed PFCs as acceptable with certain contingent restrictions. Likewise, although HFC-23 (CF_3H) has a 300-year lifetime,[15] it is a by-product of the manufacture of HCFC-22 (CF_2HCl), which will continue to be produced as an intermediate for the manufacture of polymers such as Teflon™, and EPA thus has placed no restriction on its use as a fire protection agent.

In response to concerns about environmental effects and efficacy, fire protection manufacturers are also developing several new alternative fire protection technologies, including inert gas systems, water mist systems, and powdered aerosol systems. These non-halocarbon alternative agents require a different means of determining risk during use. Some of the newer non-halocarbon alternative agents—the inert gas systems—limit but do not entirely remove the oxygen available to a fire. The most important condition for the safe use of such agents is the stipulation that the amount of oxygen remaining in the area of release is sufficient to maintain central nervous system function and that reduced oxygen does not impair escape from the area if people are exposed.

Powdered aerosol systems present still other risk assessment issues. The conditions determining the safe use of these agents must account for potential deposition in the respiratory tract of inhalable particles, ranging from very small particles that may be deposited in the alveoli to large particles capable of irritating the upper nasal passages. The size of such particles may be the most significant factor determining risk. Water mist systems using pure water pose little risk, although additives must be evaluated on a case-by-case basis to determine potential health hazards. A concern with both mist and powdered aerosol systems is the visual obscuration that occurs during discharge and that may potentially limit individuals' ability to leave the area.

Because the risk analyses for alternative fire protection technologies differ somewhat from standard EPA risk assessment procedures, EPA has encouraged the formation of ad hoc workshops and medical peer-review panels to characterize the risks presented by each new technology and to help delineate the appropriate exposure limits for different clinical groups. Conditions for the appropriate use of inert gases with limited oxygen have been evaluated by special medical panels, and EPA has also solicited guidance from OSHA on conditions of use, since OSHA will ultimately determine the proper use of all fire suppressant systems. Workshops and panels have been formed to analyze issues concerning powdered aerosols and water mists.

ALTERNATIVES TO USE OF HALONS

The EPA has been largely successful in identifying several agents and technologies that can be used in most total flooding and streaming fire protection applications. There are still some application areas that pose technical challenges, however, including aviation (both civil and military), military tanks, some military shipboard uses, and explosion inertion applications. The U.S. military has been a leader in research and development efforts, e.g., the selection of HFC-125 (CHF_2CF_3) for the design of fire protection systems on new military aircraft and the selection of HFC-227ea (CF_3-$CHFC$-CF_3; or FM-200™) for new shipboard machinery spaces. For commercial aircraft, the Federal Aviation Administration (FAA) is spearheading an industrywide R&D effort to identify effective substitutes for halon as a fire suppressant. Once an agent is identified for complex systems, much work still remains to design, manufacture, and certify the fire protection system (see Chapters 1 and 2 for details).

HALON BANG

To serve existing applications depending on equipment that cannot be retrofitted cost-effectively, the U.S. EPA encourages halon banking programs. The Department of Defense maintains such a bank for mission-critical systems, managed by the Defense Logistics Agency, which also serves as the buffer needed while new agents are identified and systems developed for new platforms. In the commercial sector, users have undertaken similar actions to redeploy and bank halon. Private-sector businesses have sprung up to work the halon recycling market, and the non-profit Halon Recycling Corporation plays an important role in aiding buyers and sellers of halon both in the United States and abroad.

The data for estimating the global supplies of halons are collected by two different methods. The first is based on amounts manufactured annually by the major producers and on emission patterns. Countries that have required collection of halon for destruction provide the second. In two such countries, Australia and Germany, the first estimates of the halon to be collected were based on the study of annual production. In both cases, these initial estimates had to be revised downward because the actual quantity of halons collected fell short of projections, possibly because (1) actual quantities within the country were less than estimated, (2) some halon was emitted rather than being collected, or (3) the halon was not turned in. Any or all these could account for the discrepancies. The major point is that we just do not know. Currently, UNEP's Halon Technical Options Committee is reexamining its estimates of the global bank of halons.

REFERENCES

1. J.D. Longstreth et al., "Effects of Increased Solar Ultraviolet Radiation on Human Health," pp. 23-48 in *Environmental Effects of Ozone Depletion: 1994 Assessment*, J.C. van der Leun, Ed., United Nations Environment Programme, Nairobi, Kenya, November (1994).
2. S. Solomon, D. Wuebbles, et al., "Ozone Depletion Potentials, Global Warming Potentials, and Future Chlorine/Bromine Loading," pp. 13.9-13.13 in *Scientific Assessment of Ozone Depletion: 1994*, Co-Chairs, Daniel L. Albritton, Robert T. Watson, and Piet J. Aucamp, World Meteorological Organization, Geneva (1994).
3. U.S. Environmental Protection Agency, *Risk Screen on the Use of Substitutes for Class I Ozone Depleting Substances: Fire Suppression and Explosion Protection*, EPA, Washington, D.C., March (1994), pp. 2-2 to 2-4.
4. Gary Taylor, "Halon Bank Management - A Rationale to Evaluate Future World Supplies," in *Proceedings of the Second International Conference on Halons and the Environment*, CFPA Europe/NFPA, Geneva, Sept. 28 (1990).
5. Environment Canada, *Halon Bank Management—A Rationale for Canada*, March 15 (1990).
6. Halon Fire Extinguishing Agents Technical Options Report to the United Nations Environment Programme Technology Review Panel, Gary Taylor and Major E. Thomas Morehouse, Jr., Co-Chairs, Toronto, Canada, June (1989), p. 9.
7. S. Solomon, D. Wuebbles, et al., "Ozone Depletion Potentials, Global Warming Potentials, and Future Chlorine/Bromine Loading," pp. 13.9-13.13 in *Scientific Assessment of Ozone Depletion: 1994*, Co-Chairs, Daniel L. Albritton, Robert T. Watson, and Piet J. Aucamp, World Meteorological Organization, Geneva (1994).
8. Ozone Secretariat, *Handbook for the Montreal Protocol on Substances That Deplete the Ozone Layer*, third edition, United Nations Environment Programme, Nairobi, Kenya, August (1993), p. 27.
9. Clean Air Act, *U.S. Code*, Vol. 42, Title VI, secs. 7450 et seq (1990).
10. Reva Rubenstein, "Human Health and Environmental Toxicity Issues for Evaluation of Halon Replacements," *Toxicology Letters* **68**, 21-24 (1993).
11. National Fire Protection Association, *Halon 1301 Fire Extinguishing Systems*, NFPA 12A, NFPA, Quincy, Massachusetts (1992), pp. 12A-11.
12. U.S. Environmental Protection Agency, Significant New Alternatives Policy Program, Rulemakings, and Notices, 59 FR 13044 (March 18, 1994); 59 FR 44240 (August 26, 1994); 60 FR 3318 (January 13, 1995).
13. Clean Air Act, *U.S. Code*, Vol. 42, Title VI, secs. 7401 et seq., sec. 612(a) , (1990).
14. A.R. Ravishankara, S. Solomon, A.A. Turnipseed, and R.F. Warren, "Atmospheric Lifetimes of Long-lived Halogenated Species," *Science* **259**, 194-199 (1993).
15. Du Pont Fluorochemicals, "EPA SNAP Submission for HFC-23," January (1993).

Appendix C

Stability and Materials Compatibility of Candidate Replacements for Halon

STORAGE STABILITY

Halon 1301 is normally stored in metal containers. It is known to be stable under these conditions for many years. Any candidate replacement must also be stable with respect to storage in metal containers for many years. The National Institute of Standards and Technology (NIST) has conducted two studies of the storage stability of candidate replacement agents. The first of these, by Peacock et al.,[1] examined the agents listed in Table C.1. Each agent was used to fill a Teflon™-lined stainless steel cylinder containing the metal coupons listed in Table C.2.

Table C.1 Agents Examined in the National Institute of Standards and Technology Study

Agent	Chemical Formula
HCFC-22	CF_2HCl
HCFC-124	CF_3-$CHFCl$
HCFC-125	CF_3-CHF_2
HFC-32/HFC-125	CF_2H_2/CF_3-CHF_2
HFC-134a	CF_3-CHF_2
FC-218	CF_3-CF_2-CF_3
HFC-227ea	CF_3-CHF-CF_3
FC-31-10	CF_3-CF_2-CF_2-CF_3
FC-116	CF_3-CF_3
FC-C-318	Cyclo-C_4F_8
HFC-236fa	CF_3-CH_2-CF_3
Halon 1301	CF_3Br
IFC-13I1	CF_3I

Table C.2 Metals Tested in the National Institute of Standards and Technology Agent Stability Study

Metal	Major Constituents	Typical Use
Nitronic-40 (21-6-9 stainless)	Cr, Mn, Ni, Fe	Pressure vessels
304 stainless	Cr, Ni, Fe	Pressure vessels and fittings
AM 355 stainless	Cr, Mo, Ni, Fe	Outlet closures
4130 alloy steel	Cr, Mn, Si, Fe	Pressure vessels
13-8 Mo alloy steel	A1, Cr, Mo, Ni, Fe	
Inconel 625	Cr, Fe, Mo, Ni	Bourdon tubes
606 1-T6 aluminum alloy	A1, Si, Fe, Mg	Compression gaskets and fittings
CDA 172 Cu-Be alloy	Cu, Be	Bourdon tubes, outlet closures

The storage cylinders were placed in 149°C ovens for 28 days. The coupons were weighed and checked for appearance before and after testing, and the agent was analyzed using Fourier transform infrared (FT-IR) spectroscopy before and after testing. Peacock et al. concluded the following:

1. "No new compounds, observable above background, were evident after 28 day, 150°C exposure of any of the candidate agents."
2. "A possible decrease in integrated areas for selected spectral peaks was observed only for CF_3I. This observation may be within experimental error, and was without accompanying formation of new compounds. An observation of a dark solid on metal coupons post-exposure for this agent may have been I_2 which could result from either degradation or impurity in the original agent. Longer term study is warranted for this agent."
3. "For the chemicals studies, stability in long-term storage and agent residue should not be major deciding factors in determining selection of appropriate agents for further study. As noted above, CF_3I could be an exception."

In a second study, Harris[2] examined FC-218, HFC-125, HFC-227ea, and CF_3I more closely, that is, over a longer time and at more than one temperature (see Table C.3).

All agents were tested in the presence of Nitronic-40, Ti-15-3-3-3, C4130, and Inconel 625. The storage vessels were stored at the prescribed temperatures for as many as 52 weeks. At specific times the cylinders were removed from storage, cooled, and the contents analyzed by FT-IR. After conducting the study, Harris concluded:

1. "The fluorocarbon agents FC-218, HFC-125, and HFC-227ea were stable at temperatures as high as 150°C for as long as 48 weeks. No by-products were formed."
2. "CF_3I degraded at 100°C and was accelerated at 150°C."
3. "CF_3H, CO_2, and CO were produced in low levels as degradation products of CF_3I."
4. "The presence of moisture accelerates the degradation of CF_3I."
5. "The presence of copper accelerates the degradation of CF_3I."
6. "The presence of copper and moisture accelerate the degradation of CF_3I."
7. "An abundance band at 950 cm^{-1} was generated in the CF_3I samples that may be from a fluorinated alkene; the presence of copper at 150°C caused the double bond to break."
8. "Storage at ambient conditions of any of the four agents is feasible, but storage at elevated temperatures for CF_3I needs more study."

Table C.3 Test Matrix for Harris Study

Agent	Temperature
FC-218	150°C
HCFC-125	23°C and 150°C
HFC-227ea	23°C, 100°C, and 150°C
CF_3I (dry)	23°C, 100°C, and 150°C
CF_3I with Cu (dry)	23°C and 150°C
CF_3I (moist)	150°C
CF_3I with Cu (moist)	150°C

EFFECTS OF HALON-LIKE ALTERNATIVE AGENTS ON ORGANIC MATERIALS AND METALS

Storage Considerations

In considering of the storage of halon-like agents, the long-term effects of the agent on elastomeric sealing gaskets and lubricants in the storage system must be considered. In this connection, a specific evaluation of various fluorocarbons (FCs), hydrofluorocarbons (HFCs), and hydrochlorofluorocarbons (HCFCs), acting on relevant elastomeric gasket materials and lubricants, has been carried out by McKenna et al.[3] at NIST. The elastomers investigated included silicones, fluorosilicones, fluorocarbons, neoprene, and nitriles. The lubricants included a fluorinated grease, a perfluoropolyether grease, and a commercial aircraft grease.

Two methods of evaluation were employed. The first method involved measurement of the degree of swelling of the elastomer or lubricant on exposure to agent vapor. If the amount of solvent absorbed was small, the agent was reckoned to have good compatibility; i.e., it did not damage the elastomer or the lubricant. Bad compatibility was implied by excessive swelling, and fair compatibility was intermediate between the two. The measurements were carried out at 35°C. This method is indirect in the sense that it does not measure the effect on the mechanical properties of the elastomer or lubricant. Good compatibility implies equilibrium sorption of a weight fraction less than 0.22, while bad compatibility implies sorption greater than 0.38.

The second method involved measurement of mechanical properties of the elastomers and lubricants after extended exposure (weeks) to agent vapor at elevated temperature and pressure (150°C, 5.86 Mpa). The authors concluded that testing at 150°C was too severe, but the data are indicative if not interpreted absolutely. For elastomers the mechanical properties are compression set resistance, i.e., the tendency to spring back after compression, and elongation reduction in ultimate tensile testing, i.e., the loss of stretchability. These mechanical factors are directly relevant to gasket performance. Bad compatibility is indicated when the ultimate elongation decreases by 65% after a 2-week exposure to agent. Fair compatibility, i.e., marginally acceptable, is indicated by a 65% loss in 4 weeks.

Lubricants responded to exposure to an agent in a different manner. The mobile fraction of the lubricant is extracted by the agent over time, leaving a powdery residue that is unsuitable as a lubricant. Bad compatibility signifies an agent that leaves the lubricant powdery after 4 weeks of exposure. Good compatibility implies that the lubricant does not become powdery after 6 weeks of exposure.

Results are given in Tables C.4 and C.5. The first letter (B, F, or G) is the poorer of the two mechanical ratings, i.e., compression set resistance or reduction of elongation. The second letter is based on swelling measurements. In drawing conclusions from the ratings one must be aware of the authors' admonition that the exposure conditions prior to mechanical testing (150°C) were too severe. Even so, until further test results are available, the information may be used for tentative screening. Examination of Table C.4 suggests that neither of the nitriles is promising as a gasket material with halon-like agents. Fluorocarbon elastomers are similarly unsuitable. Silicone, fluorosilicone, and neoprene elastomers emerge as superior candidates, but there are sufficient negative entries to make further testing necessary before selecting a material. In assessing the compatibility of lubricants, the mechanical property measure is probably most relevant. That is, the tendency to become powdery appears to be the failure mechanism, and swelling may not measure this tendency. Even so, the mechanism may be more complex, and further testing under milder conditions of exposure is indicated. The data in Table C.5 do not offer unambiguous clues for distinguishing among the three lubricants.

The study of McKenna et al. is an excellent beginning but, as the authors point out, further testing is required. Their study contains a wealth of interpretation not included herein. The results do lead to optimism that effective, long-lived gasket and lubricating materials can be found for a given halon-like fire fighting agent. It should be noted that other classes of elastomers could be considered (e.g., polyacrylates and polyphosphazenes). It is recommended that specific tests be carried out before designing a system.

Table C.4 Elastomer Compatibility

Agent	Silicone	F-silicone	Neoprene	Fluorocarbon	Nitrile	Nitrile (low T)
FC-116, C_2F_6	G/G	G/G	G/G	B/G	F/G	B/G
FC-218, C_3F_8	G/G	G/G	G/G	B/G	F/G	B/G
FC-31-10, C_4F_{10}	G/F	F/G	F/G	B/F	B/F	B/F
FC-318, c-C_4F_8	G/G	G/G	G/G	G/G	F/G	F/F
HFC-236fa, $C_3H_2F_6$	G/F	G/B	B/B	B/B	B/F	B/B
HFC-32-325*	G/F	B/G	F/G	B/G	B/G	B/G
HFC-227ea, C_3HF_7	G/F	G/F	F/G	B/F	B/G	B/B
HCF-134a, $C_2H_2F_4$	G/F	G/F	F/G	B/G	B/G	B/G
HFC-125, C_2HF_5	G/G	B/G	G/G	B/F	B/G	B/G
HCFC-22, CHF_2Cl	B/F	G/F	B/G	B/F	B/B	F/B
HCFC-124, C_2HF_4Cl	B/B	G/F	B/F	B/F	B/F	B/F

NOTE: G, F, and B mean good, fair (i.e., between good and bad) and bad, respectively. In this context, a good solvent is one that does not diminish the desired behavior of an elastomer or lubricant. Two measures are given for each, representing the separate mechanical and sorption methods of characterization. See text.

* HFC-32-325 is an azeotrope of CH_2F_2 and C_2HF_5.

Table C.5 Lubricant Compatibility

Agent	Krytox 240AC	Braycote 600	Braycote 807
FC-116, C_2F_6	F/F	G/F	F/F
FC-218, C_3F_8	B/G	B/G	B/G
FC-31-10, C_4F_{10}	B/F	B/B	B/B
FC318, c-C_4F_8	B/F	B/F	B/F
HFC-236fa, $C_3H_2F_6$	G/G	G/G	G/G
HFC-33-125*	G/G	G/G	G/G
HFC-227ea, C_3HF_7	B/F	G/F	G/F
HFC-134a, $C_2H_2F_4$	B/G	B/F	B/G
HFC-125, C_2HF_5	G/F	G/G	G/G
HCFC-22, CHF_2Cl	B/F	B/G	F/F
HCFC-124, C_2HF_4Cl	G/G	G/G	G/G

NOTE: G, F, and B mean good, fair (i.e., between good and bad) and bad, respectively. In this context, a good solvent is one that does not diminish the desired behavior of an elastomer or lubricant. Two measures are given for each, representing the separate mechanical and sorption methods of characterization. See text.

* HFC-32-125 is an azeotrope of CH_2F_2 and C_2HF_5.

The possibility of corrosion of the storage and distribution equipment induced by halon-like agents has been studied by Ricker et al.[4] as reported in the NIST document cited above. These authors document studies that indicate that the halon-like agents do not pose serious corrosion problems for metals likely to be employed. Again, specific tests are recommended before implementation of materials choices.

Effects on Plastics and Other Organic Materials

The halon-like alternative agents tend to be chemically inert under most anticipated storage and discharge conditions. They would not be expected to exhibit deleterious effects on organic materials present, for example, in equipment or in protected spaces. Examples of organic materials that may be of concern include wire insulation, packaging for solid-state circuitry, circuit boards, floor coverings, paint, furniture coatings, and so on. Machinery spaces aboard naval ships, in general, have less exposed organic material than is typical of other occupied spaces.

Owing to the short exposure time associated with agent discharge in a fire fighting incident, no significant chemical deterioration of exposed organic materials is expected. A mode of failure called environmental stress cracking does exist, and it may be prudent to conduct specific tests on relevant plastics under stress in the presence of the agent in critical areas such as wire insulation. For materials used in typical shipboard machinery spaces, the likelihood of material failure resulting from agent discharge is very low. The reader is referred to the *Modern Plastics Encyclopedia*[5] and a document from the American Society for Testing Materials[6] for more detail.

Effects of Discharged Agents on Metals

During flame extinguishment, some acid is formed by the decomposition of halons and halon-like alternative agents. This effect is smaller when the time to extinguishment is shorter. Thus, the release of a larger quantity of agent can result in a smaller quantity of acidic decomposition by-products such as HF.

REFERENCES

1. R.D. Peacock, T.G. Cleary, and R.H. Harris, Jr., pp. 643-668 in *Evaluation of Alternative In-Flight Fire Suppressants for Full-Scale Testing in Simulated Aircraft Engine Nacelles and Dry Bays*, NIST Special Pub. 861, U.S. Department of Commerce, Washington, D.C. (1994).
2. R.H. Harris, Jr., *Fire Suppression System Performance of Alternative Agents in Aircraft Engine and Dry Bay Laboratory Simulations*, Vol. 1, NIST Special Pub. 890, pp. 249-403, U.S. Department of Commerce, Washington, D.C. (1995).
3. *Evaluation of Alternative In-Flight Fire Suppressants for Full-Scale Testing in Simulated Aircraft Engine Nacelles and Dry Bays*, W.L. Grosshandler, R.G. Gann, and W.M. Pitts, Eds., pp. 729-763, U.S. Department of Commerce, Washington D.C. (1994).
4. *Evaluation of Alternative In-Flight Fire Suppressants for Full-Scale Testing in Simulated Aircraft Engine Nacelles and Dry Bays*, W.L. Grosshandler, R.G. Gann, and W.M. Pitts, Eds., pp. 669-728, U.S. Department of Commerce, Washington D.C. (1994).
5. *Modern Plastics Encyclopedia*, McGraw-Hill, New York (1996).
6. American Society for Testing Materials (ASTM), *Resistance of Plastics to Chemical Reagents*, D543, ASTM, West Conshohocken, Pa. (1995).

Appendix D

Water Mist Fire Suppression Technology

Water mist fire suppression systems have been studied for at least 50 years. While no practical or commercially demonstrated systems have been developed until recently, the basis for use of fine liquid water droplets for gas-phase fire suppression is relatively old. Recent interest in water mist technology has been driven by two events. The need for low-weight-impact replacement sprinkler systems on commercial ships driven by International Maritime Organization (IMO) regulations requiting retrofit of most commercial marine vessels gave immediate impetus to the development of low-water-demand, high-efficiency mist systems to replace sprinkler systems. The second driving force was the phase-out of halons and the search for alternative technologies that preserve all or most of the benefits of a clean total flooding agent without having a negative environmental impact.

The state of technology is such that water mist systems for replacing low-pressure water sprinkler systems aboard ships are relatively well developed and have been commercialized. The use of water mist as a replacement for halon 1301 as a total flooding agent is in its infancy but has been demonstrated for certain naval applications.

Fine water mist has been an active area of research and development, and many commercial systems are available or in development. Fine water mist relies on relatively small (<200 μm) droplet sprays to extinguish fires. In theory, the small drops allow the mist to move around obstructions and extinguish fires, mimicking characteristics of a total flooding gas. The mechanisms of extinguishment include flame cooling by droplet heating and evaporation, oxygen depletion by steam expansion, wetting of surfaces, and oxygen depletion due to combustion products.

Water mist systems may have a number of advantages, including possible low cost, absence of toxicity or adverse environmental effects, efficacy in suppression of flammable liquid pool and spray fires, and potential efficacy as inerting or explosion suppression systems. The potential efficacy of water mist fire suppression systems has been demonstrated in numerous studies and in a wide range of applications, including Class B spray and pool fires[1,2,3,4] and fires m aircraft cabins,[5,6] shipboard machinery and engine room spaces,[7,8,9,10,11,12] shipboard accommodation spaces,[13] and computer and electronics applications.[14,15]

To summarize, these experimental efforts have shown that the efficacy of a particular water mist system is strongly dependent on the ability not only to generate sufficiently small droplet sizes but also to distribute a critical concentration of droplets throughout a compartment.[16,17,18] There IS some evidence that the droplets must interact with the flame sheet with sufficient momentum to penetrate the flame. Factors that affect the distribution of this critical concentration of water mist include droplet size, velocity, the spray pattern geometry and the momentum and mixing characteristics of the spray jet, and the geometry and other characteristics of the protected area. While it is relatively easy to generate a dense aerosol of small droplets, it is more difficult to provide sufficient momentum to distribute the spray throughout the space, around obstacles, and so on. Hence, water mist must be evaluated in the context of a system, and not just as an extinguishing agent.

It is apparent that when water mist systems are being evaluated for fire extinction capability as opposed to fire suppression (an easier task), their sensitivity to the details of the area being protected must be considered. It is therefore essential to develop worst-ease fire scenarios and hazard geometries to evaluate the fire extinguishing capabilities of water mist systems.

THEORETICAL AND DESIGN CONSIDERATIONS

The major difficulties with water mist systems are those associated with design and engineering. These problems arise from the need to generate, distribute, and maintain an adequate concentration of properly sized drops throughout a compartment while gravity and agent deposition losses on surfaces deplete or reduce the concentration.

There is no current theoretical basis for predicting optimal drop size and velocity distribution, spray momentum, distribution pattern, and other important water mist system parameters. This is of course analogous to the lack of a theoretical basis for nozzle design for total flooding gaseous systems, and/or even conventional sprinkler and water spray systems.

Extensive experimental and theoretical work aimed at predicting critical adiabatic flame temperatures appears to indicate a range between 1600 to 1900 K, depending on the fuel. According to Holmstedt,[19] there are two possible methods by which water spray may extinguish a fire: by extinction of the flame or by cooling of the fuel. Holmstedt states that fuel cooling is performed by larger drop sizes and hence is only relevant to suppression of solid combustible fires. It is possible, however, to use the momentum of large droplets (> 200 μm) to drag or entrain smaller droplets, thus providing a mechanism for mixing and distribution.

The potential is present for water mist to act as a true flooding agent if the mass median drop size is below 20 microns. At this level, its suppressing efficiency is twice that of halon 1301 per unit weight. Before this becomes possible, methods of controlling the droplet transport must be developed. The amount of water required to lower the flame temperature to the range of limiting values is between 0.15 to 0.25 L/m^3 (1.0 to 1.8 gal/1000 ft^3). The actual concentration required may be less than this due to the oversimplification discussed previously.

The predominant variables contributing to the production of this concentration are drop size and flow rate. Drop size plays an important role in estimation of the required flow rate as well as in the production of a critical concentration of drops. Drops under 50 microns begin to exhibit characteristics of a gas in the increase in fall time and decrease in terminal velocity. Conversely, larger drops fall faster, resulting in greater fallout losses. Water flux densities (flow rate per unit area) vary significantly across experimental test programs, ranging from 1.5 Lpm/m^2 to as high as 10 Lpm/m^2.[20,21] The significantly higher water flux densities recommended by Gameiro and Mawhinney may be a function of inefficient production of critical concentration (i.e., greater losses due to a larger drop size, poor mixing, and so on).

The primary loss mechanism, plate loss caused by gravity and spray impact on walls and obstructions, presents a very difficult technical challenge. From a design standpoint, the loss is overcome by using larger water flow rates and continuous discharge of water. In this sense, current water mist systems greatly exceed the theoretical minimum water concentrations described previously.

Vent loss rates are a function of the size of the vent, the size of the fire (which drives the flow through the vent), any other pressure induced across a compartment boundary, and the concentration of drops in the compartment. Evaporation losses are significantly more difficult to calculate or estimate. The evaporation of a drop is a function of drop size, initial temperature, velocity with respect to the surrounding gas, gas temperature, and so on. It is worth noting, assuming all things are constant, that the life of a drop is usually proportional to the square of its diameter.

RECENT TECHNOLOGY ADVANCES

Misting/Atomization Technology

During the past decade or so, there has been an expansion of the science and technology of the transportation of bulk liquids into fine sprays (atomization). The primary contributors to this technology have been the combustion industry (fuel spray atomization), the chemical industry (spray drying), and the power industry (evaporative cooling). Significant information relevant to water mist applications for fire suppression can be extracted from this knowledge base.

Atomized sprays may be produced in various ways. Basically, all that is needed is a high relative velocity between the liquid to be atomized and the surrounding air. Some atomizers accomplish this by discharging the liquid at high velocity into a relatively slow-moving stream of air. Notable examples include the various forms of pressure atomizers. An alternative approach is to expose the relatively slow-moving liquid to a high-velocity airstream. The latter method is generally known as dual-fluid, air-assist, or air-blast atomization.

Two atomization technologies are incorporated in water mist suppression systems under development and/or consideration: single- and dual-fluid systems. Single-fluid systems (pressure atomizers) utilize water stored or pumped at high pressure (40 to 200 bar) and spray nozzles with relatively small orifice sizes. Dual-fluid systems use air, nitrogen, or other gases to atomize water at a nozzle. Both types of systems have been shown to be effective fire suppression systems.

U.S. Navy Water Mist Technology

The U.S. Navy has developed a machinery space water mist system that utilizes a modified high-pressure spray nozzle. The nozzle design is described in several Naval Research Laboratory (NRL) reports.[22,23,24] The basic design approach is to produce high volumes of 100-μm droplet (mean diameter) sprays with very high spray momentum to achieve rapid suppression of large hydrocarbon pool or spray fires. These nozzles emit 2 gpm at 1000 psi and are spaced approximately 8 ft apart on a uniform grid mounted in the overhead and at the intervening deck level in the machinery space. This system has been tested extensively on the *ex*-USS *Shadwell*, the NRL fire test vessel in Mobile Bay, Alabama. It is capable of suppressing fires in seconds. The Navy's water mist system is not particularly effective on highly obstructed small fires, although it provides substantial cooling and limits the fire size, thereby enabling relatively safe manual fire fighting.

Compared to some commercially available technology, the Navy system uses relatively high water flow application rates (approximately 0.06 to 0.07 gpm/ft^2, which is on the order of three to four times the rate of the best available systems). The relatively high water flow rate requires significant pumping and electrical power capacity. For the LPD-17's largest machinery space, a 250-hp motor is required for a 200-gpm pump.

On new-design naval vessels, electrical power and water supply are not particularly difficult constraints, and so the relative efficiency of the system is not an issue. For any retrofit application, however, the current Navy design would be problematic, and the water flow rates of this system would make the use of stored pressure cylinders (vs. pumps) quite difficult. This would substantially limit the application of this system for small individual compartments such as flammable liquid storage rooms.

One important component of the Navy system is that it was designed, developed, and tested under significant time constraints and is scheduled for installation on the LPD-17. Optimization of the system or evaluation of alternative designs can and should be pursued if additional applications, particularly retrofit or protection of small enclosures, are envisioned.

Commercial Water Mist Fire Suppression Systems

At least 11 water mist system technologies are currently available or under development using either dual-fluid (N_2/air and water) or single-fluid high-pressure systems. Table D.1 summarizes the commercially available water mist systems that can be used to protect against flammable and combustible liquid hazards. While the performance of these systems varies widely, development of this technology has just begun, and improvements in the effectiveness and efficiency of water mist systems can be expected.

Table D.1 Commercially Available Water Mist Systems

Manufacturer	Atomization Method	Pressure Source	IMO Machinery Space Approval[a]	Other Flammable/ Combustible Liquid Applications[b]
ADA Technologies/ Fike	Dual fluid atomization	Cylinders, low pressure	No	Yes
Reliable/Baumac	High pressure, single fluid	Pumps	No	Yes
Kidde International	Low pressure, single fluid	Pump	No	Yes
Grinnell	Low pressure, single fluid	Pump	No	Yes
Unifog	High pressure, single fluid	Pump/cylinders	Yes (<500 m^3)	Yes
Marioff	High pressure, single fluid	Pump or cylinders	Yes (< 3000 m^3)	Yes
Securiplex	Dual fluid, air assisted	Cylinders, low pressure	No	Yes
Wormold/Total Walther	Low pressure, single fluid	Pumps	No	Yes

[a] International Maritime Organization (IMO) approval based on successful completion of machinery space testing in accordance with MSC Circular 668 (1995) is noted along with limitations on machinery space volume for those tests.
[b] as indicated by testing performed by the U.S. Navy, U.S. Coast Guard, National Research Council of Canada, or a similar national laboratory.

EXPERIMENTAL EVALUATION OF WATER MIST SYSTEMS

The efficacy of water mist fire suppression systems recently has been demonstrated through experimental programs for a range of applications, including the following:

- Class B spray and pool fires;[25,26,27,28]
- Aircraft cabins;[29,30,31]
- Shipboard machinery and engine room spaces;[32,33,34,35,36,37]
- Shipboard accommodation spaces;[38] and
- Computer and electronics applications.[39,40]

In addition, Factory Mutual Research Corporation has developed a performance-based approval standard for water mist applications for turbine generator enclosures and machinery spaces.

The following sections partially summarize water mist system testing in applications similar to those for the U.S. Navy requiring replacement of halon 1301.

Naval Research Laboratory, Small Compartment Testing

Over 500 water mist system tests have been conducted by the NRL. Many of these tests were part of an ongoing investigation into the use of water mist as an alternative for halon in machinery space applications for the U.S. Navy.[41,42,43] These tests have included both generic systems utilizing modified industrial spray nozzles and commercially available fire protection misting hardware. The systems tested cover the spectrum of available technologies, including dual-fluid fixed orifice, dual-fluid sheet/slit orifice, single-fluid, high-pressure multiple-orifice heads, and single-fluid, high-pressure grid/matrix-type systems. It was not the intent of this set of NRL investigations to judge one system in terms of another, but rather to determine the capabilities and weaknesses of water mist technology.

Each system was evaluated in a variety of configurations to achieve optimal results. The fire-fighting capabilities of these optimized systems varied only slightly for a given flux density. The results were driven primarily by the similarity in drop size distribution between the systems, with the mass mean diameter of drops measured as $D_{v0.5}$ ~ 75 Tm ± 25 Tm. (The mass mean diameter, $D_{v0.5}$, is defined as the diameter of a drop such that 50% of the total liquid volume/mass is in drops of a smaller diameter.)

Some general observations from this effort to assess the fire-fighting performance of water mist systems are as follows:

- All of the systems evaluated were able to extinguish unobstructed fires on the floor of the compartment with spray flux densities on the order of 1.0 Lpm/m^2;
- Many fires located at higher elevations in the compartment were extinguished, and the remaining fires were reduced dramatically in size;
- Large fires are easier to extinguish than small fires owing to the displacement of oxygen by the expansion of the water mist to steam as well as to higher plume entrainment rates associated with larger fires;
- The fire-fighting capabilities of the two-fluid systems were found to increase when nitrogen and other inert gases were substituted for air as the second fluid; and
- Obstructed fires become more difficult to extinguish with increased horizontal drop travel distance (i.e., horizontal distance from the higher flux density region near the spray pattern to the fire source). Many fires were extinguished at distances on the order of 0.3 m (1 ft) but were not extinguished from greater distances. It is worth noting that many of the highly obstructed fires, although not extinguished, were greatly reduced in size by the presence of the water mist.

The MicroMist system by Baumac and the Marioff system represent the extremes of design philosophy for single-fluid, high-pressure water mist systems. One relies on spray momentum for distribution and mixing of drops; the other utilizes many nozzles that produce small droplets with virtually no spray momentum.

The major feature of the Marioff nozzle is its droplet size distribution. The flow pattern comprises both large (~100 μm) and small (<50 μm) drops. The large droplets provide spray momentum, which assists in penetration and mixing. Typical spacing is 120 to 150 ft^2 per head.

Utilizing 1000-psi water supplied by a pump, the Baumac International MicroMist system produces a large amount of very small droplets with almost no momentum in the spray. This is the system that most closely approximated a "total flooding" system. It was capable of effectively extinguishing a majority of the unobstructed fires and demonstrated superior fire-fighting capabilities (compared to the other systems tested) in the obstructed pan and comer fire scenarios. In a broader context, these extinguishment efficiencies are still dramatically lower than those of a gaseous agent and would be viewed as inadequate for a total flooding system.

Dual-fluid systems (air atomized) use air at 30 to 100 psi to atomize water supplied at 25 to 100 psi. The droplet size distribution can be varied across a wide range by changing the relative water and air flow rates, air pressure, and nozzle orifice design. Several of these types of systems are commercially available. They have been shown to be very effective against localized flammable-liquid hazards.

Table D.2 Performance of Two Navy Water Mist System Types

	Extinguishment Time (min:s)			
	Grinnell AquaMist		Spraying Systems Company	
SYSTEM CHARACTERISTICS				
Nozzle pressure	18 bar (250 psi)	18 bar (250 psi)	105 bar (1500 psi)	105 bar (1500 psi)
System flow rate	310 Lpm (82 gpm)	310 Lpm (82 gpm)	166 Lpm (44 gpm)	166 Lpm (44 gpm)
Application rate	1.86 Lpm/m^2 (0.046 gpm/ft^2)	3.69 Lpm/m^2 (0.091 gpm/ft^2)	1.01 Lpm/m^2 (0.025 gpm/ft^2)	2.03 Lpm/m^2 (0.050 gpm/ft^2)
Nozzle location	Single level	Bilevel	Single level	Bilevel
SCENARIO				
Scenario 2 (4.5 MW)*				
Fire #1	3:30	1:18	1:45	0:20
Fire #2	4:00	0:41	1:27	0:20
Fire #3	0:50	No	0:40	0:20
Fire #5	2:05	No	1:30	0:20
Scenario 4 (7.5 MW)*				
Fire #1	1:40	2:05	1:20	0:20
Fire #2	2:25	1:02	1:25	0:15
Fire #3	0:20	1:12	0:50	0:20
Fire #5	0:55	No	1:15	0:25
Scenario 5 (7.5 MW)**				
Fire #1	3:41	No	2:30	0:40
Fire #2	No	No	No	No
Fire #3	0:55	No	0:20	0:40
Fire #5	No	2:40	1:15	0:50

* Ventilation (exhaust and supply) was secured during mist system activation.
** Ventilation (exhaust and supply) remained operating during this test.

Naval Research Laboratory, Machinery Space Testing

The NRL conducted several series of full-scale tests aboard the *ex* -USS *Shadwell*, a damage control and fire-fighting test vessel located in Mobile Bay, Alabama. Testing was performed in a simulated machinery space with a volume of 926 m^3 (36,000 ft^3). Five water mist nozzle types were evaluated. The tests, which were conducted m 1994 and 1995,[44,45,46,47] included diesel and heptane pressurized spray and pool fires and heptane pool fires ranging in size from 3.5 to 7.5 MW. Table D.2 summarizes the performance of two systems, including the Spraying Systems Company's modified nozzle for a two-level nozzle grid installation. For all fires, the extinction times with the high-pressure system were very short. The only failure to extinguish occurred in Scenario 5, where the ventilation system was not left operating throughout the tests. Scenario 4 was identical to Scenario 5 except that ventilation was secured. In an actual installation, the ventilation would be secured, as it is for halon systems, during the system discharge.

U.S. Coast Guard, Machinery Space Testing

The U.S. Coast Guard conducted a series of full-scale tests in a simulated 560-m^3 machinery space. The tests were designed to evaluate the performance of five different water mist nozzles, including the performance of the Spraying Systems Company/U.S. Navy nozzle as measured by the IMO test procedure for merchant vessel machinery space fire protection systems. The IMO test procedure includes heptane and diesel fuel spray and pool fires ranging in size from 1.0 to 6.0 MW.

Overall, these tests demonstrated the ability of water mist systems to control or extinguish a range of liquid pool and spray fires. The tests underscored the difficulty of extinguishing small or obstructed fires and the importance of compartment size in evaluating mist systems (due to oxygen depletion caused by the fire). The tests also demonstrated the generally superior performance of high-pressure (>60 bar) nozzles, including the Navy nozzle, particularly in extinguishing small or obstructed fires.

Factory Mutual Gas Turbine Enclosure Testing

Factory Mutual Research Corporation (FMRC) has developed a fire test procedure for evaluating water mist-based fire protection systems for combustion turbine enclosures. At least three commercial water mist systems have successfully completed the testing regime, but only one system (Securiplex) has completed all product certification requirements.

The FMRC test procedure is used to evaluate water mist systems for compartments up to 260 m^3. It is designed to demonstrate system performance on relatively small (1 MW), highly obstructed, flammable and combustible liquid pool and spray fires. An additional important feature of the testing is evaluation of the heat transfer rates between the mist spray and turbine casings. Tests are conducted and heat transfer modeling performed to ensure that, under worst-case conditions, the turbine casing does not deflect excessively.

While the FMRC tests are conducted in compartments that are relatively small compared to Navy machinery spaces, the results are of interest for naval gas turbine enclosures and also are indicative oft he performance of current commercial water mist hardware. These systems are significantly more efficient than the Navy mist system owing to several factors. Since the use of pumped systems has significant cost and complexity penalties for commercial applications, stored-pressure systems are preferred. As a result, very space- and weight-efficient, completely self-contained water mist systems have been developed. For example, one system utilizes three 50-L cylinders and four nozzles to provide protection for 15 minutes in an enclosure with a volume of 260 m^3, a capability that may be of interest for the protection of Navy flammable storage spaces and turbine enclosures. Another feature of these systems is that the water flow is cycled on and off, a technique that not only minimizes water use but also has been shown to improve fire extinguishing performance.

The FMRC approval testing has demonstrated the efficiency of water mist in this application. The systems developed are very cost effective and have performance advantages over halon 1301 and other total flooding gas systems.

National Research Council of Canada

Mawhinney[48] developed engineering design criteria for machinery space water mist fire suppression systems based on Canadian Navy experiments conducted at the National Fire Laboratory in Canada. Key characteristics of such systems include drop size distribution, spray flux, and spray momentum, among others. If the spray is mixed with additives, this, too, is an important characteristic that can affect performance. Test results are based on the use of a particular set of nozzles for a particular set of conditions. Committing to one particular nozzle or system design for all applications would not be appropriate until all engineering constraints have been analyzed. System design must be based on fire suppression objectives and overall system economics in making the decisions on whether to use low-pressure, intermediate-pressure, high-pressure, or twin fluid nozzles.

Water mist does have total flooding limitations, particularly for the nozzles tested by Mawhinney, as do all types of total flooding fire suppression systems. For the purposes of the Canadian Navy, the maximum compartment size is set at 200 m^3 to meet economic and space and weight restrictions for storage.

Mawhinney concluded that water mist is potentially an effective fire suppressant for hydrocarbon liquid pool and spray fires depending on the geometry of the compartment.

Sintef (Norway)

A fire research organization in Norway (Sintef) has conducted extensive tests on water mist in flammable and combustible liquid fires. The tests were performed on two different scales. The first was a 30-m^3 test enclosure used to develop the characteristics for the full-scale 70-m^3 enclosure. The purpose of the phase I work was to identify the extinguishing characteristics of various BP Sunbury Research Center nobles and determine the efficiency of Ginge-Kerr Offshore's total fire suppression system. The suppression system had a dual fluid nozzle design using air and water at 5 bar. The nozzles produced a high-velocity, small-droplet water mist.

Phase II tested and evaluated the efficiency of fine water spray nozzles in fighting various turbine hood fires in a full-scale test enclosure, consisting of an engine mock-up used to simulate the hot engine surfaces, insulation mats, and piping that would be found in a real engine hood. Diesel pool and spray fires, and diesel-soaked insulation mat fires, were fought under differing conditions of air flow and nozzle position and flow.

The test results ran the full range of possibilities. Large underventilated gas, pool, and oil spray fires were extinguished with the addition of small amounts of water. This was due to near self-extinguishment caused by lack of oxygen being introduced into the hood.

Large well-ventilated gas, pool, and oil spray fires, and fires from oil spray hitting hot metal surfaces, produced varying results. The fires were extinguished in the cases where the mist was able to reach the base of the fire, but not when the droplets could not do so. The oil spray fire on hot metal surfaces was extinguished consistently when the water spray system covered the full area at which the oil spray hit the metal surface, even in the cases when the metal surface temperature remained high.

It was found that 1-m^2 (medium) well-ventilated pool fires, small pool fires ($\ll 1\ m^2$), and fires in oil-soaked insulation mats were very difficult to extinguish. The droplets were not able to penetrate the fire to effectively evaporate the water in the flame zone, nor could they reach the base of the fire.

In the final condition, oil-soaked insulation mats with hot metal surfaces below the mat, the fires were extinguished successfully but had a tendency to reignite. Reignition could be curbed with sustained addition of the water mist to both displace oxygen and cool the metal surface.

The effectiveness of water in the form of a fine water spray as an extinguishant has been demonstrated recently in full-scale testing conducted at Sintef laboratories in Trondheim, Norway. A full-scale mock-up of an enclosed ABB Stal GT-35 gas turbine was used for the purpose of these tests.

In conclusion, the ability of a fine water spray to extinguish fires in gas turbines has met the initial performance requirements with substantial safety margins built in. In installations equipped with 200 liters of water, only 10 liters were required to extinguish a large fire, leaving ample amounts for additional discharges. The concerns of thermal shock were resolved.

CURRENT STATUS OF WATER MIST SYSTEMS

The efficacy of water mist fire suppression systems as an alternative to halon 1301 or other total flooding gases in naval and marine flammable and combustible liquid hazard areas, including machinery spaces, has been demonstrated. Water mist has a particular advantage over gases due to the substantial environmental and hot-surface cooling that occurs. While water mist systems may also have advantages with respect to reduced space and weight requirements and lower cost relative to total flooding gas systems, these parameters vary widely for systems and specific applications.

The Navy-developed nozzle, a modification of a commercially available high-pressure nozzle, has demonstrated good performance over a range of fire scenarios and may represent substantial improvements in water mist technology relative to the other designs currently contemplated for use by the Navy. Such improvements may provide opportunities to broaden the application of water mist systems through improvements in space and weight impacts and retrofit potential.

However, the need for substantial full-scale testing to improve the currently inadequate fundamental understanding of water mist fire suppression and extinguishment mechanisms represents a substantial cost, timing, and optimization barrier to additional development.

REFERENCES

1. P.G. Papavergos, "Fine Water Sprays for Fire Protection," *Proceedings of the Halon Alternatives Technical Conference*, Albuquerque, N. Mex., May (1991).
2. J.R. Butz and R. Carey, "Application of Fine Water Mists to Fire Suppression," *Proceedings of the Halon Alternatives Technical Conference* , Albuquerque, N. Mex., May (1991).
3. R. Wighus, "Fine Water Spray Against Hydrocarbon Fires," SINTEF — Norwegian Fire Research Laboratory, Trondheim, Norway (1993); R. Wighus, "Fine Water Spray System — Extinguishing Tests in Medium and Full Scale Turbine Hood," Norwegian Fire Research Laboratory (1993).
4. C.S. Cousin, "Recent Work on Fire Control Using Fine Water Sprays at the Fire Research Station," *Proceedings of the First International Conference on Fire Suppression Research*, Stockholm and Bors, Sweden, May 5-8 (1992).
5. R.G. Hill, C.P. Sarkos, and T.R. Marker, "Development and Evaluation of an On-Board Aircraft Water Spray System for Postcrash Fire Protection," SAE Technical Paper 912224, Aerospace Technology Conference and Exposition, Long Beach, California, September 23-26 (1991).
6. R.T. Whitfield, Q.A. Whitfield, and J. Steel, "Aircraft Cabin Fire Suppression by Means of an Interior Water Spray System," CAA Paper 88014, Civil Aviation Authority, July (1988).
7. J.R. Mawhinney, "Fine Water Spray Fire Suppression Project," *Proceedings of the First International Conference on Fire Suppression Research* , Stockholm and Bors, Sweden, May 5-8 (1992).
8. A.R.F. Turner, "Water Mist in Marine Applications," presented at the Water Mist Fire Suppression Workshop, National Institute of Standards and Technology, Gaithersburg, Maryland, March 1 (1993).
9. M. Arvindson and A. Ryderman, "Tests in Simulated Ship's Engine Rooms with Hi-fog Fire Protection System," 91 R30189, Swedish National Testing and Research Institute, Bors, Sweden, July 28 (1992).
10. M. Tuomissari, "Fire Suppression Tests in Simulated Ship's Engine Room with a Hi-fog Fire Protection System," PAL 2210/92, VTT Fire Technology Laboratory, Helsinki, Finland, November 16 (1992); M. Tuomissari, "Enclosed Space Fire Suppression Tests," PAL 2206/92, October 23 (1992); M. Tuomissari, "Extinguishing Tests of Simulated Computer Room Fires by a Hi-fog Sprinkler System," PAL 2196/92, August 11 (1992); and M. Tuomissari "Withstand Voltage of Switch Gears in the Presence of Operating Hi-fog Fire Protection System," 9AFX92-98, ABB Strömberg Research Centre, Vassa, Finland, August 3 (1992).
11. E. Soja, "DGME Waterfog Trials," YARD Report No. 4175-NM0609, British Ministry of Defense - Navy, Bath, England (1990).
12. V. Gameiro, "Fine Water Spray Technology Water Mist Fire Suppression Systems," presented at the Halon Alternatives Technical Working Conference, Albuquerque, N. Mex., May 11-13 (1993); V. Gameiro, "Fine Water Spray Fire Suppression Alternative to Halon 1301 in Machinery Spaces," *The 1993 International CFC and Halon Alternatives Conference Proceedings* , Washington, D.C., October 20-22 (1993).
13. M. Arvindson and A. Ryderman, "Tests in Simulated Ship's Engine Rooms with Hi-fog Fire Protection System," 91 R30189, Swedish National Testing and Research Institute, Bors, Sweden, July 28 (1992).
14. A.T. Hills, T. Simpson, and D.P. Smith, "Water Mist Fire Protection Systems for Telecommunications Switch Gear and Other Electronic Facilities," presented at the Water Mist Fire Suppression Workshop, National Institute of Standards and Technology, Gaithersburg, Maryland, March 1 (1993).
15. M. Tuomissari, "Fire Suppression Tests in Simulated Ship's Engine Room with a Hi-fog Fire Protection System," PAL 2210/92, VTT Fire Technology Laboratory, Helsinki, Finland, November 16 (1992); M. Tuomissari, "Enclosed Space Fire Suppression Tests," PAL 2206/92, October 23 (1992); M. Tuomissari, "Extinguishing Tests of Simulated Computer Room Fires by a Hi-fog Sprinkler System," PAL 2196/92, August 11 (1992); and M. Tuomissari "Withstand Voltage of Switch Gears in the Presence of Operating Hi-fog Fire Protection System," 9AFX92-98, ABB Strömberg Research Centre, Vassa, Finland, August 3 (1992).

16. J.R. Mawhinney, "Design of Water Mist Fire Suppression Systems for Shipboard Enclosures," *Water Mist Instead of Halon? — International Conference on Water Mist Fire Suppression Systems*, Borås, Sweden, November 4-5 (1993); J.R. Mawhinney, "Waterfog Fire Suppression System Project: Full Scale Fire Tests Summary Report," ND Project No. DNASE40291, National Research Council Canada, Ottawa (1993); J.R. Mawhinney, "Engineering Criteria for Water Mist Fire Suppression Systems," presented at the Water Mist Fire Suppression Workshop, National Institute of Standards and Technology, Gaithersburg, Maryland, March 1 (1993); and J.R. Mawhinney, "Characteristics of Water Mists for Fire Suppression in Enclosures," *Halon Alternatives Technical Working Conference 1993*, New Mexico Engineering Research Institute, Albuquerque, N. Mex. (1993).
17. C.S. Cousin, "Recent Work on Fire Control Using Fine Water Sprays at the Fire Research Station," *Proceedings of the First International Conference on Fire Suppression Research*, Stockholm and Bors, Sweden, May 5-8 (1992).
18. L.A. Jackman, "Mathematical Model of the Interaction of Sprinkler Spray Drops with Fire Gases," *Proceedings of the First International Conference on Fire Suppression Research*, Stockholm and Bors, Sweden, May 5-8 (1992).
19. G. Holmstedt, "Extinction Mechanisms of Water Mist," *Water Mist Instead of Halon? — International Conference on Water Mist Fire Suppression Systems*, Borås, Sweden, November 4-5 (1993).
20. V. Gameiro, "Fine Water Spray Technology Water Mist Fire Suppression Systems," presented at the Halon Alternatives Technical Working Conference, Albuquerque, N. Mex., May 11-13 (1993); and V. Gameiro, "Fine Water Spray Fire Suppression Alternative to Halon 1301 in Machinery Spaces," *The 1993 International CFC and Halon Alternatives Conference Proceedings* , Washington, D.C., October 20-22 (1993).
21. J.R. Mawhinney, "Design of Water Mist Fire Suppression Systems for Shipboard Enclosures," *Water Mist Instead of Halon? — International Conference on Water Mist Fire Suppression Systems*, Borås, Sweden, November 4-5 (1993); and J.R. Mawhinney, "Waterfog Fire Suppression System Project: Full Scale Fire Tests Summary Report," ND Project No. DNASE40291, National Research Council Canada, Ottawa (1993); J.R. Mawhinney, "Engineering Criteria for Water Mist Fire Suppression Systems," presented at the Water Mist Fire Suppression Workshop, National Institute of Standards and Technology, Gaithersburg, Maryland, March 1 (1993); and J.R. Mawhinney, "Characteristics of Water Mists for Fire Suppression in Enclosures," *Halon Alternatives Technical Working Conference 1993*, New Mexico Engineering Research Institute, Albuquerque, N. Mex. (1993).
22. J.T. Leonard et al., "Full-scale Machinery Space Water Mist Tests: Phase I — Unobstructed Space," NRL Ltr. Rpt. Ser. 6180/0713.1, Naval Research Laboratory, Washington, D.C., October (1994).
23. J.T. Leonard et al., "Full Scale Machinery Space Water Mist Tests: Phase II — Simulated Machinery Spaces," NRL Ltr. Rpt. Ser. 6180/0868.2, Naval Research Laboratory, Washington, D.C., December (1994).
24. J.T. Leonard, G.G. Back, P.J. DiNenno, and R.L. Darwin, "Full-scale Tests of Water Mist Fire Suppression Systems for Navy Shipboard Machinery Spaces: Part II — Obstructed Spaces," NRL/MR/6180-96-7831, Naval Research Laboratory, Washington, D.C., March 8 (1996).
25. P.G. Papavergos, "Fine Water Sprays for Fire Protection," *Proceedings of the Halon Alternatives Technical Conference*, Albuquerque, N. Mex., May (1991).
26. J.R. Butz and R. Carey, "Application of Fine Water Mists to Fire Suppression," *Proceedings of the Halon Alternatives Technical Conference* , Albuquerque, N. Mex., May (1991).
27. R. Wighus, "Extinguishment of Enclosed Gas Fires with Water Sprays," *Fire Safety Science — Proceedings of the Third International Symposium* , Edinburgh, (Elsevier, ISBN 1-85166-719-9) (1991); and R. Wighus, "Active Fire Protection — Extinguishment of Enclosed Gas Fires with Water Sprays," SINTEF Report STF25 A91028, Trondheim, Norway (1991).
28. C.S. Cousin, "Recent Work on Fire Control Using Fine Water Sprays at the Fire Research Station," *Proceedings of the First International Conference on Fire Suppression Research*, Stockholm and Bors, Sweden, May 5-8 (1992).
29. R.G. Hill, C.P. Sarkos, and T.R. Marker, "Development and Evaluation of an On-Board Aircraft Water Spray System for Postcrash Fire Protection," SAE Technical Paper 912224, Aerospace Technology Conference and Exposition, Long Beach, California, September 23-26 (1991).
30. R.G. Hill, T.R. Marker, and C.P. Sarkos, "Evaluation of an On-Board Water Spray Fire Suppression System in Aircraft," presented at the Water Mist Fire Suppression Workshop, National Institute of Standards and Technology, Gaithersburg, Maryland, March 1 (1993).
31. R.T. Whitfield, Q.A. Whitfield, and J. Steel, "Aircraft Cabin Fire Suppression by Means of an Interior Water Spray System," CAA Paper 88014, Civil Aviation Authority, July (1988).

32. J.R. Mawhinney, "Fine Water Spray Fire Suppression Project," *Proceedings of the First International Conference on Fire Suppression Research*, Stockholm and Bors, Sweden, May 5-8 (1992).
33. E. Soja, "DGME Waterfog Trials," YARD Report No. 4175-NM0609, British Ministry of Defense - Navy, Bath, England (1990).
34. V. Gameiro, "Fine Water Spray Technology Water Mist Fire Suppression Systems," presented at the Halon Alternatives Technical Working Conference, Albuquerque, N. Mex., May 11-13 (1993); V. Gameiro, "Fine Water Spray Fire Suppression Alternative to Halon 1301 in Machinery Spaces," *The 1993 International CFC and Halon Alternatives Conference Proceedings* , Washington, D.C., October 20-22 (1993).
35. A.R.F. Turner, "Water Mist in Marine Applications," presented at the Water Mist Fire Suppression Workshop , National Institute of Standards and Technology, Gaithersburg, Maryland, March 1 (1993).
36. M. Arvindson and A. Ryderman, "Tests in Simulated Ship's Engine Rooms with Hi-fog Fire Protection System," 91 R30189, Swedish National Testing and Research Institute, Bors, Sweden, July 28 (1992).
37. M. Tuomissari, "Fire Suppression Tests in Simulated Ship's Engine Room with a Hi-fog Fire Protection System," PAL 2210/92, VTT Fire Technology Laboratory, Helsinki, Finland, November 16 (1992); M. Tuomissari, "Enclosed Space Fire Suppression Tests," PAL 2206/92, October 23 (1992); M. Tuomissari, "Extinguishing Tests of Simulated Computer Room Fires by a Hi-fog Sprinkler System," PAL 2196/92, August 11 (1992); and M. Tuomissari "Withstand Voltage of Switch Gears in the Presence of Operating Hi-fog Fire Protection System," 9AFX92-98, ABB Strömberg Research Centre, Vassa, Finland, August 3 (1992).
38. M. Arvindson and A. Ryderman, "Tests in Simulated Ship's Engine Rooms with Hi-fog Fire Protection System," 91 R30189, Swedish National Testing and Research Institute, Bors, Sweden, July 28 (1992).
39. R.G. Hill, T.R. Marker, and C.P. Sarkos, "Evaluation of an On-Board Water Spray Fire Suppression System in Aircraft," presented at the Water Mist Fire Suppression Workshop, National Institute of Standards and Technology, Gaithersburg, Maryland, March 1 (1993).
40. M. Tuomissari, "Fire Suppression Tests in Simulated Ship's Engine Room with a Hi-fog Fire Protection System," PAL 2210/92, VTT Fire Technology Laboratory, Helsinki, Finland, November 16 (1992); M. Tuomissari, "Enclosed Space Fire Suppression Tests," PAL 2206/92, October 23 (1992); M. Tuomissari, "Extinguishing Tests of Simulated Computer Room Fires by a Hi-fog Sprinkler System," PAL 2196/92, August 11 (1992); and M. Tuomissari "Withstand Voltage of Switch Gears in the Presence of Operating Hi-fog Fire Protection System," 9AFX92-98, ABB Strömberg Research Centre, Vassa, Finland, August 3 (1992).
41. J.T. Leonard et al., "Full-scale Machinery Space Water Mist Tests: Phase I — Unobstructed Space," NRL Ltr. Rpt. Ser. 6180/0713.1, Naval Research Laboratory, Washington, D.C., October (1994).
42. J.T. Leonard et al., "Full Scale Machinery Space Water Mist Tests: Phase II — Simulated Machinery Spaces," NRL Ltr. Rpt. Ser. 6180/0868.2, Naval Research Laboratory, Washington, D.C., December (1994).
43. J.T. Leonard, G.G. Back, P.J. DiNenno, and R.L. Darwin, "Full-scale Tests of Water Mist Fire Suppression Systems for Navy Shipboard Machinery Spaces: Part II — Obstructed Spaces," NRL/MR/6180-96-7831, Naval Research Laboratory, Washington, D.C., March 8 (1996).
44. J.T. Leonard et al., "Full-scale Machinery Space Water Mist Tests: Phase I — Unobstructed Space," NRL Ltr. Rpt. Ser. 6180/0713.1, Naval Research Laboratory, Washington, D.C., October (1994).
45. J.T. Leonard et al., "Full Scale Machinery Space Water Mist Tests: Phase II — Simulated Machinery Spaces," NRL Ltr. Rpt. Ser. 6180/0868.2, Naval Research Laboratory, Washington, D.C., December (1994).
46. J.T. Leonard, G.G. Back, P.J. DiNenno, and R.L. Darwin, "Full-scale Tests of Water Mist Fire Suppression Systems for Navy Shipboard Machinery Spaces: Part II — Obstructed Spaces," NRL/MR/6180-96-7831, Naval Research Laboratory, Washington, D.C., March 8 (1996).
47. G.G. Back et al., "Full-scale Testing of Water Mist Fire Suppression Systems in Machinery Spaces," U.S. Department of Transportation, U.S. Coast Guard, Groton, Conn., November (1995); M. Tuomissari, "Suppression of Compartment Fires with a Small Amount of Water," *Water Mist Instead of Halon? — International Conference on Water Mist Fire Suppression Systems*, Borås, Sweden, November 4-5 (1993).
48. J.R. Mawhinney, "Fine Water Spray Fire Suppression Project," *Proceedings of the First International Conference on Fire Suppression Research*, Stockholm and Bors, Sweden, May 5-8 (1992).

Appendix E

Glossary

Activation energy: the energy, over and above the ground state, which must be added to a reaction system in order to prompt the reaction to occur.

Adiabatic: referring to any change in which there is no gain or loss of heat.

AFFF: aqueous film-forming foam; a water-based foam used primarily to fight flight-deck fires aboard ships that carry aircraft.

Alkanes: hydrocarbons having the formula C_nH_{2n+2}.

Atmospheric lifetime: average length of time a compound survives in the atmosphere.

Bilge: the inside bottom of a ship or boat.

CAAA: Clean Air Act Amendments of 1990.

Cardiotoxicity: toxic effects with impact on heart tissue or function.

CFCs: chlorofluorocarbons; carbon compounds containing both chlorine and fluorine.

CVN: nuclear-powered aircraft carrier.

DDG: a guided-missile destroyer.

DNA: deoxyribonucleic acid; the molecule that stores genetic information in biological systems.

EKG: electrocardiogram

Endothermic reaction: a chemical reaction that absorbs heat.

Enthalpy: also known as heat content or total heat; the sum of the internal energy of a system plus the product of its volume multiplied by the pressure exerted on the system by its surroundings.

Entropy: a measure of the disorder of a system, equal to the Boltzmann constant times the natural logarithm of the number of microscopic states corresponding to the thermodynamic state of the system.

EPA: Environmental Protection Agency.

FAA: Federal Aviation Administration.

Fetotoxicity: toxic effects with impact on a fetus.

FFG: Fiscal and Force Capability Guidance guided-missile frigate.

Frigate: an ocean escort ship that is smaller and slower than a destroyer.

Greenhouse gas: a gas that absorbs infrared radiation and thus contributes to radiative forcing.

GWP: global warming potential; an index that represents a cumulative measure of the expected effects on global climate from the emissions of a gas relative to carbon dioxide (CO_2).

Halocarbon: a carbon compound containing one or more of the halogen elements (fluorine, chlorine, bromine, and iodine).

Halon: or halons; a generic term referring to halocarbons containing bromine that are widely used as fire and explosion suppressants. The term "halon" is used in this report to refer to halon 1301 (CF_3Br) and halon 1211 (CF_2ClBr).

HARC: Halon Alternatives Research Corporation.

HCFCs: hydrochlorofluorocarbons; hydrocarbons that contain chlorine and fluorine.

Heat capacity: the quantity of heat required to raise the temperature of a unit mass of a substance by a fixed amount. Materials with high heat capacity can absorb heat efficiently.

Heat of formation: the increase in enthalpy (total heat) resulting from the formation of a fixed quantity of material from its constituent elements.

Heterogeneous reaction: a chemical reaction system in which the reacting agents are in different physical phases. For example, one reactant may be a gas and the other a liquid.

HFCs: hydrofluorocarbons; hydrocarbons that contain fluorine.

HRC: Halon Recycling Corporation.

HTOC: Halon Technical Options Committee.

IMO: International Maritime Organization

Inert gas generators: devices containing a rapidly combustible material that yields chemically inert gases, such as nitrogen, upon combustion. Inert gas generators are widely used in automotive air bags and are used in some fire suppression systems where the inert gas displaces oxygen and thus inhibits combustion.

Inertion: interruption or suppression of an explosion.

Infrared radiation: electromagnetic radiation with wavelengths in the range of 0.8 to 1000 µm, which is longer than those of visible light and shorter than those of microwaves.

JSF: Joint Strike Fighter.

Lagging: the insulation around pipes aboard a ship.

Latent heat: the amount of heat absorbed or evolved by a unit mass of a substance during a phase change (such as melting, sublimation, or vaporization).

LCAC: landing craft, air cushion.

LHA: a general-purpose amphibious assault ship.

LHD: a multipurpose amphibious assault ship.

LOAEL: lowest observable adverse effect level.

LPD: an amphibious transport dock ship.

Mesosphere: the region of the atmosphere extending from about 45 to 55 km to about 80 to 95 km, characterized by a general decrease in temperature with increasing altitude. The mesosphere extends from the top of the stratosphere to the temperature minimum (mesopause) at the interface between the mesosphere and the thermosphere.

MSC: Military Sealift Command.

Nacelle: a separate streamlined enclosure on an airplane for sheltering or housing something, such as the crew or an engine.

NASA: National Aeronautics and Space Administration.

NAVAIR: Naval Air Systems Command.

NAVSEA: Naval Sea Systems Command.

NFPA: National Fire Protection Association.

NIST: National Institute of Standards and Technology.

NOAA: National Oceanic and Atmospheric Administration.

NOAEL: no observable adverse effect level.

ODP: ozone depletion potential; an index that represents a cumulative measure of the expected effects on ozone from the emissions of a gas relative to CFC-11 ($CFCl_3$).

OECD: Organisation for Economic Cooperation and Development, an association of the world's wealthiest and most industrialized nations that, among other things, publishes guidelines on various technical subjects related to regulations, standards, and testing.

OSHA: Occupational Safety and Health Administration.

Ozone: a reactive oxygen compound (O_3) found in the atmosphere; it shields Earth's surface from UV radiation.

Ozone layer: that part of the atmosphere in which the concentration of ozone is greatest. The term is used in two ways: (1) to signify the layer from about 10 to 50 km in which the ozone concentration is appreciable, and (2) to signify the much narrower region from about 20 to 25 km in which the concentration reaches a maximum.

PFCs: perfluorocarbons; hydrocarbons in which all of the hydrogen atoms have been replaced by fluorine.

Phase change: the metamorphosis of a material from one physical phase to another, such as gas to liquid, liquid to solid, or solid to gas.

Photodissociation: synonymous with photolysis.

Photolysis: chemical decomposition induced by the absorption of light.

PKP: potassium bicarbonate powder.

Radiative forcing: net change in irradiance (in watts per meter squared, Wm^{-2}) at the tropopause after allowing for stratospheric temperatures to readjust to equilibrium.

Radical: or free radical; an atom or molecule that has one unpaired electron.

RDT&E: research, deployment, test, and evaluation.

Retrofit: installation of new or updated systems in existing platforms.

SNAP: Significant New Alternatives Program.

Stratosphere: that region of the atmosphere lying above the tropopause in which, in contrast to adjoining regions, temperature does not increase with increasing height. The stratosphere extends from the tropopause to a height of about 50 km, where the temperature reaches a maximum.

Streaming agent: a fire suppression agent (typically liquid) applied in a stream, often from hand-held extinguishers, directly on an unobstructed fire.

Teraflop: 1 million million (10^{12}) floating point operations per second.

TFA: trifluoroacetic acid (CF_3COOH).

Thermosphere: the atmospheric shell extending from the top of the mesosphere to outer space. It is a region of more or less steadily increasing temperature with increasing altitude, starting at about 80 to 95 km.

Total flooding agent: an easily dispersed fire suppression agent (typically gaseous) for use in obstructed spaces.

Tropodegradable: tending to degrade, or break down into constituent elements, in the lower atmosphere (troposphere). Tropodegradable compounds emitted at the surface are not likely to be transported to the stratosphere.

Tropopause: the atmospheric boundary between the troposphere and the stratosphere, defined as the level at which the decrease in temperature with height becomes 2°C/km or less, over at least 2 km.

Troposphere: the lowest layer of the atmosphere below the tropopause extending from Earth's surface to about 8 km at the poles and 16 km at the equator.

Ullage: space above fuel inside a partially filled fuel tank.

Ultraviolet (UV) radiation: electromagnetic radiation lying beyond the violet end of the visible spectrum with wavelength less than 400 nm.

UNEP: United Nations Environment Programme.

UV-B radiation: electromagnetic radiation with wavelength in the range of 280 to 320 nm, which affects plants and animals.

V-22: "Osprey"; a vertical takeoff and landing tilt-rotor aircraft currently under development.

Volume mixing ratio: the ratio of the number of molecules of a species in air to the total number of gas molecules in a sample.

Water mist systems: fire suppression systems that dispense water in the form of finely aerosolized droplets. The fine water mist behaves, in some instances, like a gaseous agent and is able to be dispersed effectively in obstructed spaces.

Window region: a region in the infrared portion of the electromagnetic spectrum with wavelengths in the range of 8 to 12 μm. The atmosphere is relatively transparent to radiation in this spectral region.

WMO: World Meteorological Organization.